Certificate Mechanical Engineering Science

The Technical College Series

edited by

E. G. Sterland, J.P., M.A., B.Sc.(Eng.), C.Eng., F.I.Mech.E., F.R.Ae.S.
Principal, Rolls-Royce Technical College, Bristol

Applied Mechanics
J. D. Walker, B.Sc.(Eng.), C.Eng., F.I.Mech.E.

Certificate Mathematics Vol. 1
N. Abbott, B.Sc.(Tech.), A.F.I.M.A., Dip.Math.

Introduction to Workshop Technology
M. H. A. Kempster, C.Eng., M.I.Mech.E., A.F.R.Ae.S., M.I.Prod.E.

also of interest

Manufacturing Technology
M. Haslehurst, D.I.C., C.Eng., M.I.Mech.E., M.I.Prod.E.

Mechanical Technology Vol. 1
K. S. Ashford, B.Sc.(Eng.), C.Eng., M.I.Mech.E.
J. O. Mann, M.Sc., C.Eng., F.I.Mech.E., F.I.Nuc.E.

Materials for the Engineering Technician
R. A. Higgins, B.Sc.(Birm.), F.I.M.

Certificate Mechanical Engineering Science

J. D. Walker
B.Sc.(Eng.), C.Eng., F.I.Mech.E.
Head of the Department of Mechanical and Production Engineering
The Polytechnic, Huddersfield

The English Universities Press Ltd

ISBN 0 340 14952 3 (paper)
0 340 15240 0 (boards)

First published (as *National Certificate Mechanical Engineering Science*) 1965
Reprinted 1967
Second edition 1973

The English Universities Press Ltd
St Paul's House Warwick Lane London EC4P 4AH

Printed and bound in Great Britain by
Hazell Watson & Viney Ltd Aylesbury Bucks

To G and M

General Editor's Foreword

The Technical College Series covers a wide range of technician and craft courses, and includes books designed to cover subjects in National Certificate and Diploma courses and City and Guilds Technician and Craft syllabuses. This important sector of technical education has been the subject of very considerable changes over the past few years. The more recent of these have been the result of the establishment of the Training Boards, under the Industrial Training Act. Although the Boards have no direct responsibility for education, their activities in ensuring proper training in industry have had a marked influence on the complementary courses which Technical Colleges must provide. For example, the introduction of the module system of training for craftsmen by the Engineering Industry Training Board led directly to the City and Guilds 500 series of courses.

The Haslegrave Committee on Technician Courses and Examinations reported late in 1969, and made recommendations for far-reaching administrative changes which will undoubtedly eventually result in new syllabuses and examination requirements.

It should, perhaps, be emphasised that these changes are being made not for their own sake, but to meet the needs of industry and the young men and women who are seeking to equip themselves for a career in industry. And industry and technology are changing at an unprecedented rate, so that technical education must be more concerned with fundamental principles than with techniques.

Many of the books in the Technical College Series are now standard works, having stood the test of time over a long period of years. Such books are reviewed from time to time and new editions published to keep them up to date, both in respect of new technological developments and changing examination requirements. For instance, these books have had to be rewritten in the metric system, using SI units. To keep pace with the rapid changes taking place both in courses and in technology, new works are constantly being added to the list. The publishers are fully aware of the part that well-written up-to-date textbooks can play in supplementing teaching, and it is their intention that the Technical College Series shall continue to make a substantial contribution to the development of technical education.

E. G. STERLAND

Preface

This book has been written to help students in their Mechanical Engineering Science studies in the Ordinary National Certificate and Diploma courses in Engineering or Technology.

To present the fundemental principles of mechanical engineering science in a manner interesting and acceptable to the student; to furnish him with a liberal quantity of worked examples, including graphical solutions; to supply him with a number of exercises for his own practice—these are the aims which the author has had predominantly in mind during the preparation of this work.

The introduction of SI units has helped to resolve many of the difficulties which arose from the different systems of units previously in use, particularly in the dynamics section of the subject. Undoubtedly the student will benefit from the rational system now being accepted, and particularly by the fact that it will become standard practice.

Typical questions have been taken from the examination papers set by the following organisations: the Union of Lancashire and Cheshire Institutes, the East Midlands Educational Institution, the Union of Educational Institutions, the Northern Counties Technical Examinations Council, the Yorkshire Council for Further Education, the Welsh Joint Education Committee. In many cases the numerical content of the question has been converted into SI units; these conversions are the author's responsibility and do not necessarily mean that the units chosen would be the choice of the respective examining bodies, whose cooperation in this matter is appreciated by the author.

The author would like to place on record his appreciation of the valuable help given to him by the publishers and the general editor, and his indebtedness to many of his colleagues for their help in the work which conversion to SI units has involved and also to all those who have kindly written to point out an error or to indicate an appreciation.

J. D. WALKER

Contents

A note for the student

It is a sound principle in life never to begin anything with an apology.

It is, however, a fact that you have reached this stage in your engineering studies by one of at least two routes. You may have been in the General Engineering Course and your textbook may have been *Engineering Science*—a book in this series. In this case, some of the matter, diagrams and problems in this present book will be familiar to you, since you have seen them in the earlier book. On the other hand, you may have come to this course direct from GCE studies.

If the matter is already familiar to you—then turn over to the new work.

If the matter is not familiar to you—then you in fact are the student for whom it has been repeated.

Chapter One
Units

The Système International d'Unités, abbreviated to SI, has been accepted by the Internal Organisation for Standardisation as the legal standard. It is being introduced all over the world and will ultimately be the only system of units in use in technical and business areas of work, and also in everyday life.

The system is based on six units:

Quantity	*Unit*	*Symbol*
length	metre	m
mass	kilogram	kg
time	second	s
electric current	ampere	A
temperature	kelvin	K
luminous intensity	candela	cd

Other units and combinations of units are, with limited exceptions, made up from these.

Multiples of these units are expressed, again with limited exceptions, in powers of 1000, i.e. 10^3, 10^6, 10^{-3}, 10^{-6}, and are given characteristic names of which the following are the more usual

Name	*Symbol*	*Multiply by*	
mega	M	1 000 000	10^6
kilo	k	1 000	10^3
milli	m	0·001	10^{-3}
micro	μ	0·000 001	10^{-6}

The following units are used in this book:

Length	metre	m
	kilometre	km
	millimetre	mm
	centimetre	cm
Area	square metre	m^2
	square millimetre	mm^2
Volume	cubic metre	m^3
	litre (1000 cm^3)	l
Mass	kilogram	kg
	gram	g
	tonne (1000 kg)	t
Density	kilogram per cubic metre	kg/m^3

Quantity	Unit	Symbol
Time	second	s
	minute	min
	hour	h
Velocity	kilometre per hour	km/h
	metre per second	m/s
Acceleration	metres per second per second	m/s^2
Temperature	kelvin	K
	degree Celsius	°C
Force	newton	N
	kilonewton	kN
Torque and moment	newton metre	Nm
	kilonewton metre	kNm
Work and energy	joule	J
	kilojoule	kJ
Power	watt	W
	kilowatt	kW
Stress	newton per square metre	N/m^2
	newton per square millimetre	N/mm^2
	kilonewton per square millimetre	kN/mm^2
Pressure	newton per square metre	N/m^2
	bar (10^5 N/m^2)	

Chapter Two
Vectors

A vector quantity is one which can be completely defined only by stating the direction in which it acts, as well as its magnitude.

Many quantities can be defined simply in terms of magnitude. For example, we may indicate the quantity of bolts in a storage bin as 150, their length as 40 mm, the angle of the thread as 60° and the temperature of the room as 20° C. All these quantities, and many others, are known as scalar quantities.

In stating that the movement of an aeroplane is 500 km, we have omitted an important part of the definition by not stating the direction of the movement or displacement. The full statement '500 km due east' includes both magnitude and direction.

We can state the size or magnitude of a force as 50 units, but, in order to define the force completely, we must know the direction in which it is acting.

Quantities which are defined by magnitude only are known as scalar quantities.

Quantities which are defined by magnitude and direction are known as vector quantities.

A vector quantity is indicated by either 50_{60° or 50 N 30° E. The subscript 60° in the first case indicates an angle of 60° measured anticlockwise from the X direction, i.e. east (Fig. 2.1).

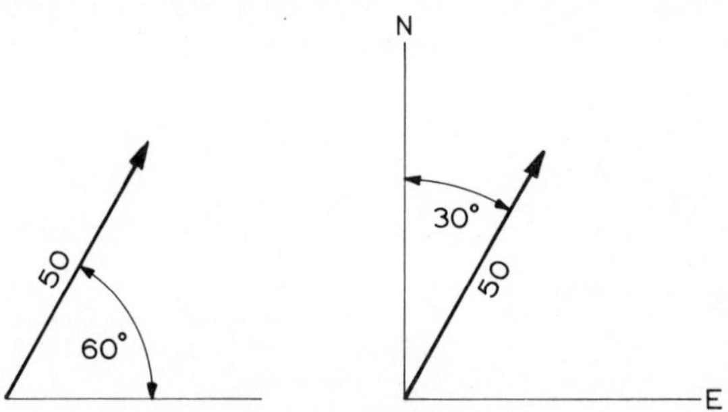

Figure 2·1

Vector quantities include forces, displacements, velocities and accelerations. They can be presented graphically by a line in which

(*a*) the length of line indicates the magnitude,
(*b*) the angle of line represents the direction, and
(*c*) an arrow on the line indicates the sense of the vector (Fig. 2.2).

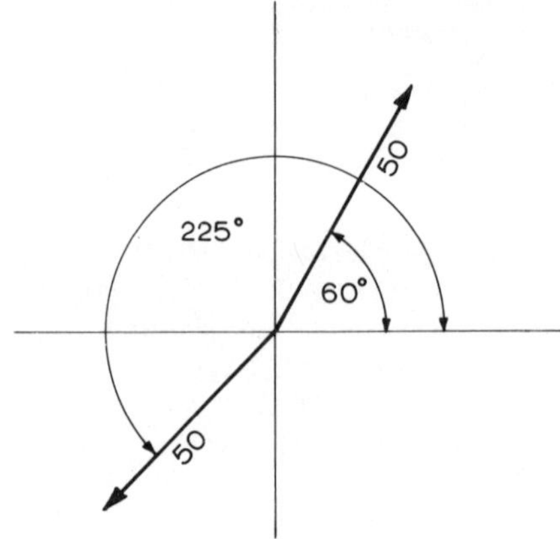

Figure 2·2

This line is called a vector and is indicated by bold type ***oa*** and the convention is that the direction (sense) is from *o* to *a*. Sometimes an arrow is placed over *oa* printed in italics thus, $\overrightarrow{oa}$.

The importance of the directional aspect of the vector quantity is illustrated in the example shown in Fig. 2.3. Two forces, one of magnitude 30 units, the other of 40 units, are shown acting in the same direction in (i) and in opposite directions in (ii). In the first case, the result of these two forces is a force of 70 units in the same direction. In the second case, the resultant of the two forces is a force of 10 units in the direction of the 40 unit force. Clearly it is not sufficient to state that two forces of given magnitudes act at a point; the resultant effect depends upon the direction of the forces.

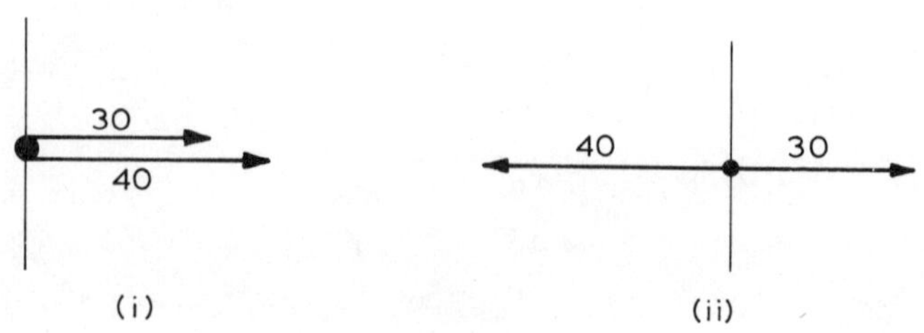

Figure 2·3

Now suppose that the direction of the 30 unit force was inclined to that of the 40 unit force, e.g. in some direction between the two extremes shown in Fig. 2.3. Suppose it was at right angles to the 40 unit force as shown in Fig. 2.4; what would be the resultant effect of these two forces?

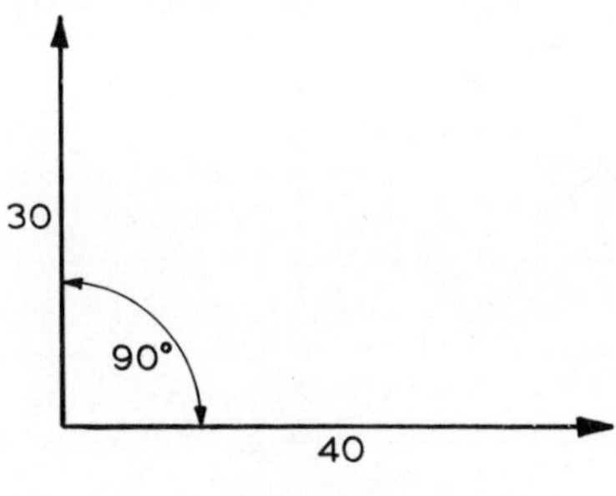

Figure 2·4

This can be determined by adding the two forces together vectorially. In fact this is what was done in Fig. 2.3, although it was done by adding +40 to −30 and getting the answer +10.

The addition of vectors not in line is carried out in the following manner (Fig. 2.5):

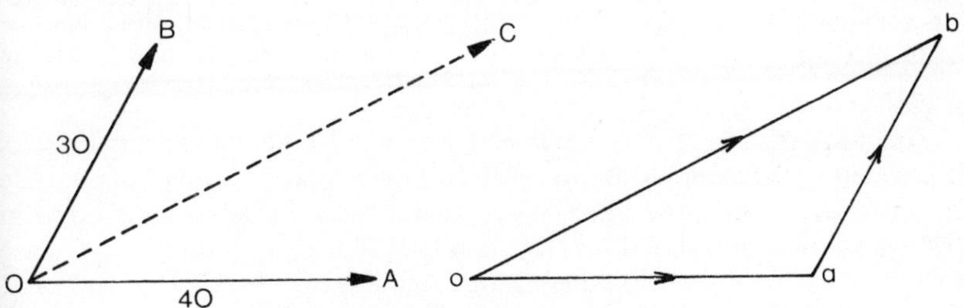

Figure 2.5 *Addition of Vectors*

(1) Draw a diagram showing the directions of the forces *OA* and *OB*.
(2) Draw vector ***oa*** to represent force *OA* in magnitude and direction.
(3) From *a*, draw vector ***ab*** to represent force *OB* in magnitude and direction.
(4) Join *ob*, then ***ob*** is the sum of ***oa*** and ***ab*** and represents in magnitude and direction the sum of the two forces *OA* and *OB*.
Note that the direction is always from *o*, in this case towards *b*.
The direction of the final vector is *always* from the *start*, *o*, to the *finish*.

Other examples in which the 30 unit force has a different inclination are shown in Fig. 2.6. In each case, the dotted line shows the sum or resultant of the two forces.

Sum of a number of vectors

If the sum of a number of vectors is required (Fig. 2.7), the process described above is extended by drawing vector ***bc*** to represent the third force; ***oc*** is then the sum of the three vectors and represents the sum of the

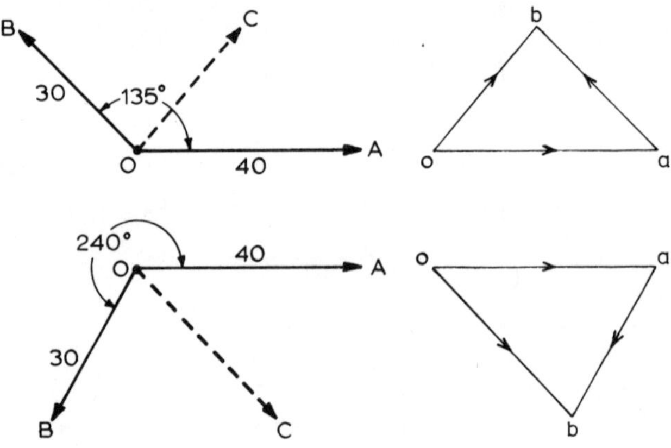

Figure 2·6

three forces in magnitude and direction. This process can be extended to any number of forces whose sum is required.

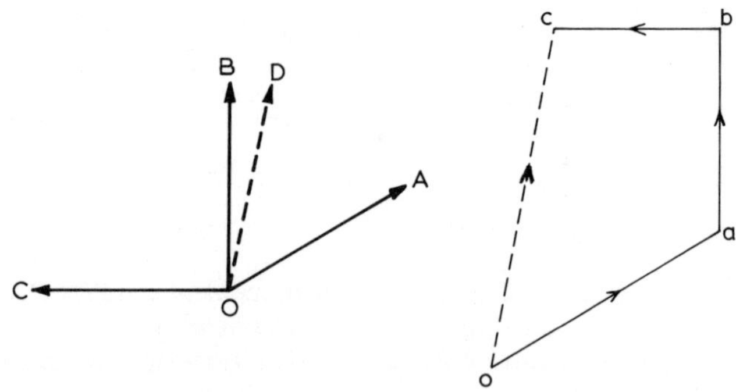

Figure 2.7 *Addition of Vectors*

Difference of two vectors

The difference of two quantities A and B can be expressed as the sum of one of the quantities and the negative of the other.

$$A - B = A + (-B)$$

We obtain the difference between the two vectors ***oa*** and ***ob*** by adding one of them to the negative of the other (Fig. 2.8).

Resolution of vectors

We have seen the method of combining two vectors to make a single vector. The reverse process of replacing one vector by two vectors is a

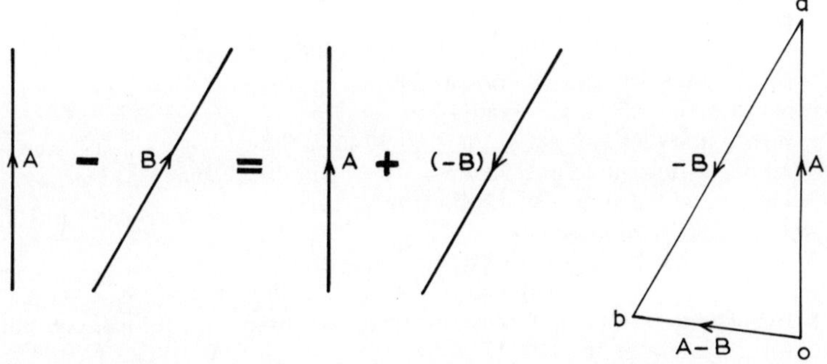

Figure 2.8 *Difference of Vectors*

very important technique in engineering science. It is known as resolving a vector into components. We are generally concerned with two components, although the process is not necessarily limited to any two components.

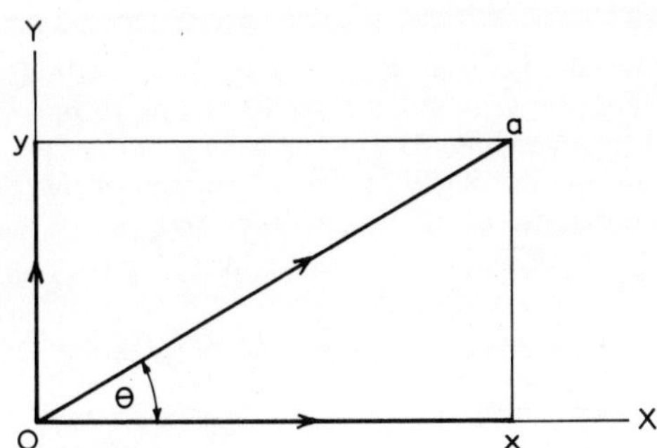

Figure 2.9 *Resolution of Vectors into Components*

The simplest case is that of the components of a vector in the horizontal and vertical directions. Take ***oa*** as an example in Fig. 2.9. A perpendicular drawn from *a* on to OX gives the point *x* and a similar perpendicular from *a* on to OY gives the point *y*.

Then ***ox*** and ***oy*** are the components of ***oa*** in the OX and OY directions.

$$\boldsymbol{ox} + \boldsymbol{oy} = \boldsymbol{oa}$$

Note that the horizontal component ***ox*** is equal to ***oa*** $\cos \theta$ and the vertical component ***oy*** is equal to ***oa*** $\sin \theta$.

EXERCISE 2

1 Determine the resultant of the following forces:
(*a*) 100 units south and 60 units north-east,

(*b*) 50 units east and 50 units south-west.

2 Find the resultant of the forces in the following cases.
(*a*) 20 units 30° and 30 units at 240°,
(*b*) 20 units 120° and 30 units 210°,
(*c*) 20 units 60°, 30 units 270° and 20 units 300°,
all angles measured anticlockwise from OX direction.

3 Determine the sum of the following vectors:
$120_{45°}$, $80_{150°}$ and $100_{210°}$.

4 What is the value of (*a*) $100_{45°} - 80_{60°}$,
(*b*) $50_{210°} - 40_{300°}$?

5 Resolve the following forces into their horizontal and vertical components.
(*a*) 200 units at 30°,
(*b*) 50 units at 45°,
(*c*) 100 units at 90°,
(*d*) 80 units at 150°,
(*e*) 40 units at 240°
(*f*) 100 units at 300°.

6 A ship sails 50 km due east and then 40 km south-east. What is its total displacement?

7 Determine the sum of the following vectors; $6_{60°}$, $10_{210°}$ and $8_{315°}$.

8 What vector must be added to those in the last question to make the sum equal to zero?

9 A force of 250 units is supplied to a support in a wall. The force makes an angle of 30° with the wall. What force is tending to pull the support out of the wall? What force is tending to bend the support?

10 *oabc* is a parallelogram. Show that ***ob*** is the sum of the vectors ***oa*** and ***oc***, and ***ac*** is the difference of the vectors, (***oc*** - ***oa***).

Chapter Three
Velocity and Acceleration

I suppose that you will all have seen a speedometer working, either on a car or bus or motor-cycle or even on an ordinary bicycle. The pointer moves to the figure 60 and we say that the car is travelling at a speed of 60 kilometres per hour. What do we actually mean by this statement? I am sure that you will see the flaw in the argument put forward by the lady when she was challenged by the police patrol for driving a car at a speed of over 60 kilometres per hour. She said that this was impossible, since she had only been away from home for 10 minutes. You see, a car has neither to travel through a distance of 60 kilometres nor for a time interval of 1 hour to be moving at a speed of 60 kilometres per hour. It does mean that if you kept the speed of the car constant, in 1 hour you would travel a distance of 60 kilometres.

Speed involves both distance and time. We can use any units to measure the distance and the time interval. If you stick a piece of paper over the 60 on the speedometer and write 1000 on the paper, this figure will indicate the number of metres through which the car will travel in 1 minute. Therefore, if you travel farther than 1000 metres in one minute you are exceeding a speed limit, which happens to have been stated in kilometres and hour units.

Hence, 60 kilometres per hour, and
16·67 metres per second, and
1000 metres per minute

are all different ways of measuring and expressing the same speed. The figures tell us that at the end of 1 second the distance travelled is increased by 16·67 metres. We are told the rate at which the distance is increasing.

DEFINITION. **Speed is the rate of increase of distance.**

Speed and velocity

When we speak of speed we are concerned only with the number value or the magnitude. When we speak of velocity we shall be concerned with both the magnitude and the direction in which the body is moving.

For example: the speed of the aircraft was 1000 kilometres per hour, but the velocity of the aircraft was 1000 kilometres per hour due north.

Velocity is a vector quantity involving magnitude and direction.

Speed is a scalar quantity involving magnitude only.

This difference between speed and velocity is most clearly shown when we consider changes in speed and changes in velocity. The speed of the

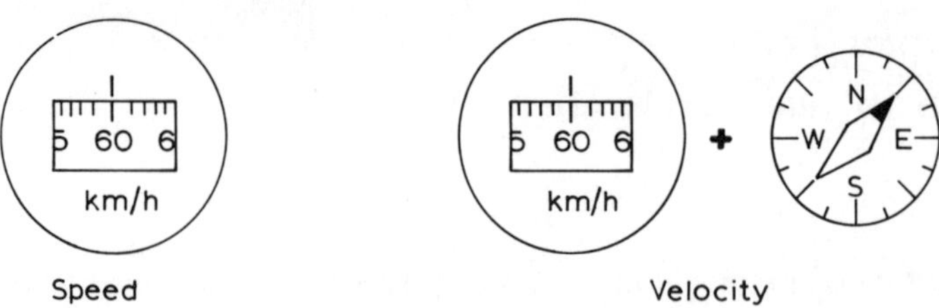

Figure 3.1 *Speed and Velocity*

Speed deals with the magnitude or number part. Velocity has to do with both the magnitude and the direction of the motion.

aircraft is changed if the number of kilometres per hour is changed, e.g. if the speedometer reads 1000 km/h then 1100 km/h. The velocity of the aircraft is changed if either:

(*a*) the number of kilometres per hour is changed, i.e. the speedometer reads 1000 km/h then 1100 km/h, or
(*b*) the direction of flight is changed, i.e. from 1000 km/h due north to 1000 km/h due east.

If you travel round a corner in a car and notice that the speedometer readings remain constant, then—

the speed is constant; but
the velocity changes, since the direction of the car is changing.

You will probably be thinking that the difference is not important. Later on, you will realise the great importance there is in this idea of velocity changing when direction changes.

Useful conversion factor

$$\begin{aligned} 1 \text{ kilometre per hour} &= 1000 \text{ metres per hour} \\ &= \frac{1000}{60 \times 60} \text{ metres per second} \\ &= \tfrac{10}{36} = 0{\cdot}28 \text{ m/s.} \end{aligned}$$

Acceleration

Whenever the velocity of a body changes, we say that the body is accelerating.

Usually acceleration refers to an increase in velocity and retardation to a decrease in velocity. We shall find it convenient to think of a positive acceleration indicating an increase in velocity and a negative acceleration indicating a decrease in velocity.

Imagine that you are in a car fitted with a speedometer indicating speeds in metres per second, and that you also have a stop-watch. Now write down the speed of the car at the end of every second. You will have to get someone to help you if you are to get some accurate figures. Your list of readings may be like this one:

End of 1st second	4 m/s
End of 2nd second	8 m/s
End of 3rd second	12 m/s
End of 4th second	16 m/s

Well, obviously the car is accelerating, its speed is increasing. The speed is increasing uniformly by 4 m/s for every second. We say that the speed is increasing at the rate of 4 m per second per second. Alternatively, we say that the acceleration of the car is 4 m per second per second, and we mean that at the end of every second the speed of the car has increased by 4 m per second. This is *uniform acceleration.*

We shall abbreviate the expression 4 m per second per second to 4 m/s^2.

DEFINITION. **Acceleration is the rate of change of velocity.**

Distance-time graph

We can plot the variation of distance with time on a graph known as a distance-time graph. Figs. 3.2 and 3.3 indicate the motion of two bodies.

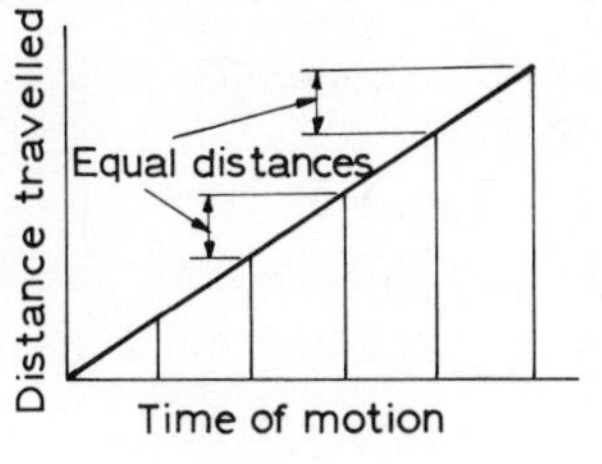

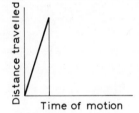

Figure 3.2 Figure 3.3

Before we compare the two motions, consider the graph on the left. The distance travelled by the body in the first second is the same as that travelled in the fourth second, in fact in any other second. In other words the body is travelling with a *constant velocity.* The distance-time graph of a body moving with a uniform velocity is always a straight line.

Now notice that in the case of both the left- and the right-hand diagrams, the bodies cover the same distance. But the one shown in Fig. 3.2 takes five times as long as the one shown in Fig. 3·3. You will notice that the graph in Fig. 3.3 is much steeper than that in Fig. 3.2. This is important in our study of motion. A steep distance-time graph indicates a high velocity. A more gradual distance-time graph indicates a low velocity. Fig. 3.4 shows a general distance-time graph of the kind you may like to plot next time you are out in a car. Read the distance travelled on the recorder at regular intervals of time and then plot the resulting graph. If the graph is steep, you have been travelling fast. If the graph is fairly level your speed is low. When the graph is horizontal you have stopped.

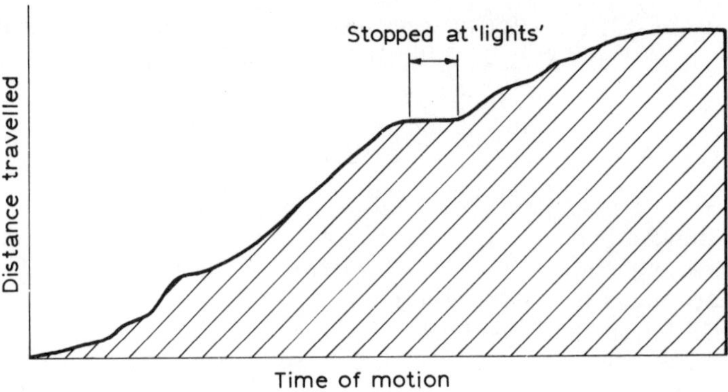

Figure 3.4 *Distance-time Graph*

Velocity-time graph

If a body has a constant velocity of 10 metres per second, it will have travelled—

10 m at the end of the 1st second
20 m at the end of the 2nd second
30 m at the end of the 3rd second,
and so on, and
$10t$ m at the end of t seconds

so that moving with a uniform or constant velocity v m/s, the body will travel a distance of vt m in t seconds.

Distance = uniform velocity × time.

A graph of velocity plotted against time will be a straight line as in Fig. 3.5.

The horizontal line represents v m/s.

The vertical line is drawn at t seconds.

The area of the rectangle is vt square units. This is numerically equal to the distance travelled by the body in t seconds.

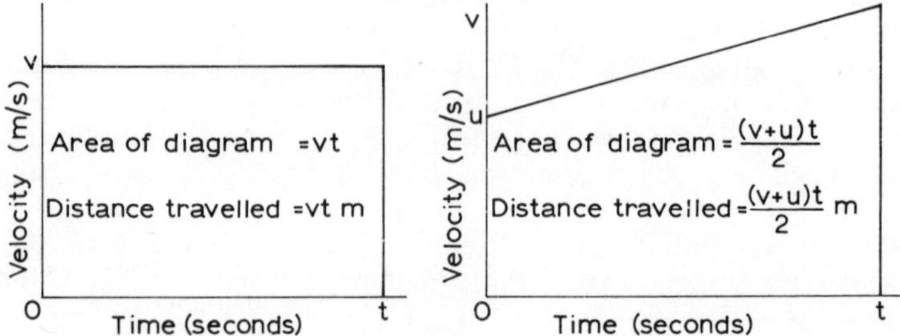

Figure 3.5 *Velocity-time Graph for Constant Velocity*

Figure 3.6 *Velocity-time Graph for Constant Acceleration*

Suppose that the velocity increases uniformly from u m/s to v m/s in the time t seconds. Now in this case the average velocity is

$$\frac{u + v}{2}\text{m/s.}$$

$$\begin{aligned}\text{Distance travelled} &= \text{average velocity} \times \text{time}\\ &= \frac{u + v}{2} \times t\\ &= \frac{(u + v)t}{2}\ \text{m.}\end{aligned}$$

The velocity-time graph is shown in Fig. 3.6.
The area of the trapezium is:

$$\begin{aligned}&\tfrac{1}{2}\text{ sum of parallel sides} \times \text{distance between them}\\ &= \tfrac{1}{2}(u + v) \times t\\ &= \frac{(u + v)t}{2}\ \text{square units.}\end{aligned}$$

This is numerically equal to the expression for the distance travelled.

In all cases, however complicated the changes in velocity may be, the area under the velocity-time graph is always numerically equal to the distance travelled by the body.

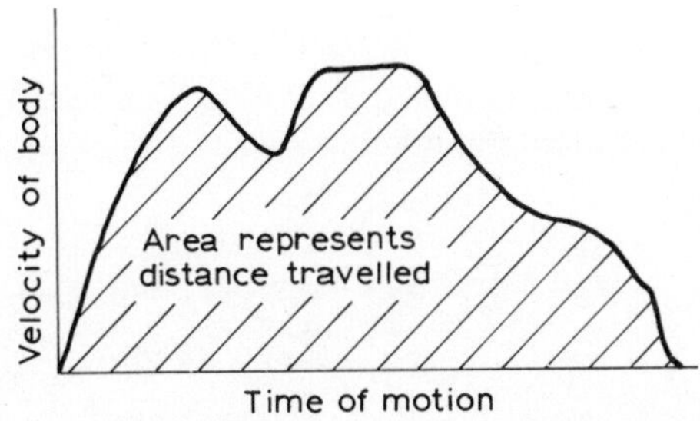

Figure 3.7 *Velocity-time Graph*

However complicated is the motion of a body, the actual distance travelled can be obtained by measuring the area under the velocity-time graph.

NOTE. If you refer to Appendix 1, you will find some methods of obtaining the area under the curve.

Distance, velocity, acceleration and time

Now let us establish some expressions which connect these four important quantities. The following symbols are usually adopted:

u = initial velocity in m/s
v = final velocity in m/s after t seconds
f = acceleration in m/s^2 (a is also used)
s = distance travelled in metres
t = time in seconds.

We are only concerned, for the present, with uniformly accelerated motion, i.e. motion in which the velocity is increasing or decreasing uniformly.

There are many cases in practice where the acceleration varies. We shall have to discover methods, at a later stage, to deal with this type of motion.

The initial velocity u will be changed to

$u + f$ at the end of 1 second
$u + 2f$ at the end of 2 seconds
$u + 3f$ at the end of 3 seconds
$u + ft$ at the end of t seconds.

Representing this by v, the final velocity, we have

$$v = u + ft \tag{1}$$

Taking u across to the other side, we have

$$v - u = ft \tag{2}$$

Now we have seen that

$$s = \frac{(u + v)t}{2} \tag{3}$$

which can easily be converted into

$$u + v = \frac{2s}{t}$$

or

$$v + u = \frac{2s}{t} \tag{4}$$

Multiply equations (2) and (4) together

$$(v - u)(v + u) = \frac{2s}{t} \times ft$$

The left-hand side is the difference of two squares, and on the right-hand side we can cancel out the 't's.' Hence we get

$$v^2 - u^2 = 2fs \tag{5}$$

Again substitute the value of v as given in equation (1) into equation (3):

$$s = \frac{[u + (u + ft)]t}{2}$$

$$s = \frac{(2u + ft)t}{2}$$

$$s = ut + \tfrac{1}{2}ft^2 \tag{6}$$

These formulae are extremely important. For convenience they are collected together here.

Uniform velocity ($f = o$)

$$s = vt$$

Uniform acceleration

$$v = u + ft$$

$$s = \frac{(u + v)t}{2}$$

$$v^2 - u^2 = 2fs$$

$$s = ut + \tfrac{1}{2}ft^2$$

If the body is accelerating, f will be positive.
If the body is retarding, f will be negative.

Example 3.1

A motor-car is travelling at a constant speed of 30 kilometres per hour. How far will the car go in 15 min? How long will it take the car to travel 10 kilometres?

If the speed is constant, distance = speed × time

kilometres = km/h × hours

$= 30 \times \frac{15}{60}$

$= 7\tfrac{1}{2}$ kilometres

$$\text{Time (hours)} = \frac{\text{distance (kilometres)}}{\text{speed (km/h)}}$$

$= \frac{10}{30}$

$= \frac{1}{3}$ hour, or 20 min.

The car will travel $7\tfrac{1}{2}$ kilometres in 15 min, and will take 20 min to travel 10 kilometres.

Example 3.2

At a given instant, the velocity of a body is 10 m/s, and it is moving with a constant acceleration of 4 m/s^2.

(i) How far will the body travel in the next 5 seconds?

(ii) What will then be its velocity?
(iii) How far will it have travelled when its velocity is 12 m/s?

(i) Initial velocity $u = 10$ m/s
Acceleration $f = 4$ m/s²
Time $t = 5$ seconds
Distance travelled $s = ?$
Now
$$s = ut + \tfrac{1}{2}ft^2$$
$$s = (10 \times 5) + (\tfrac{1}{2} \times 4 \times 5^2)$$
$$s = 100 \text{ m.}$$

(ii) Initial velocity $u = 10$ m/s
Acceleration $f = 4$ m/s²
Time $t = 5$ seconds
Final velocity $v = ?$
Now
$$v = u + ft$$
$$v = 10 + (4 \times 5)$$
$$v = 30 \text{ m/s.}$$

(iii) Initial velocity $u = 10$ m/s
Final velocity $v = 12$ m/s
Acceleration $f = 4$ m/s²
Distance $s = ?$
Now
$$v^2 - u^2 = 2fs$$
$$s = \frac{v^2 - u^2}{2f}$$
$$s = \frac{12^2 - 10^2}{2 \times 4}$$
$$s = \tfrac{44}{8}$$
$$s = 5\tfrac{1}{2} \text{ m.}$$

So the body will travel 100 m in the 5 seconds, will have a velocity of 30 m/s at the end of the 5 seconds, and will have travelled $5\frac{1}{2}$ m when its velocity is 12 m/s.

Example 3.3

A body having an initial velocity of 10 m/s and moving with a uniform acceleration covers a distance of 5500 m in 50 seconds. Calculate the acceleration and the maximum velocity.

Initial velocity $u = 10$ m/s
Final velocity $v = ?$ m/s
Distance travelled $s = 5500$ m
Time taken $t = 50$ seconds
Acceleration $f = ?$
$$s = ut + \tfrac{1}{2}ft^2$$
$$5500 = (10 \times 50) + (\tfrac{1}{2} \times 50^2 f)$$
$$f = \frac{5000 \times 2}{50^2}$$
$$f = 4 \text{ m/s}^2$$
Also
$$v = u + ft$$

$$v = 10 + (4 \times 50)$$
$$v = 210 \text{ m/s}$$

The acceleration is 4 m/s² and the maximum velocity is 210 m/s.

Example 3.4

A car at a given instant has a speed of 20 metres per second. It is given a uniform retardation of 2·5 m/s² until its speed is reduced to 10 metres per second. Determine the time taken in reducing the speed and the distance in metres the car has travelled during the retardation.

Time taken to reduce speed:

$$u = 20 \text{ m/s}$$
$$v = 10 \text{ m/s}$$
$$f = -2{\cdot}5 \text{ m/s}^2 \text{ since car is retarding}$$
$$v = u + ft$$
$$10 = 20 - 2{\cdot}5t$$
$$t = \frac{10}{2{\cdot}5} \text{ seconds}$$
$$= 4 \text{ seconds.}$$

Distance travelled during retardation:

$$v^2 - u^2 = 2fs$$
$$10^2 - 20^2 = 2 \times (-2{\cdot}5)s$$
$$s = \frac{300}{5}$$
$$= 60 \text{ m.}$$

The car takes 4 seconds to slow down, during which time it travels 60 metres.

Alternative method

The distance can be found by using average velocity. If the retardation is uniform, then the average velocity is $\frac{v + u}{2}$ and

$$\text{Distance} = \text{average velocity} \times \text{time}$$
$$= \frac{10 + 20}{2} \times 4$$
$$= 15 \times 4$$
$$= 60 \text{ metres.}$$

Example 3.5

A lift rises 100 m in 12 seconds. For the first quarter of the distance the velocity is uniformly accelerated, and during the last quarter it is uniformly retarded. If the velocity is constant during the centre portion of the ascent, determine its value.

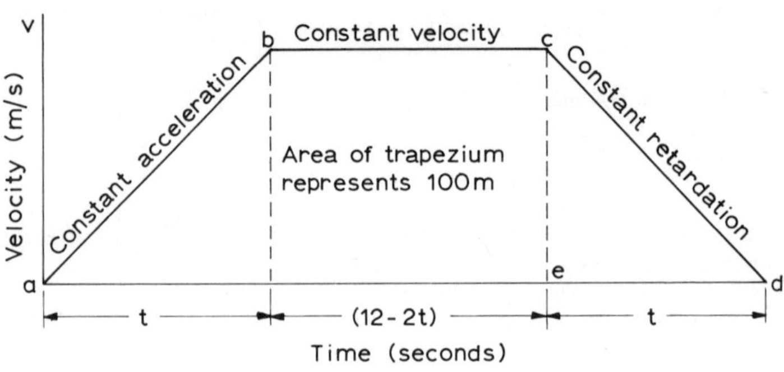

Figure 3.8

The velocity time graph is shown in Fig. 3.8.

Let v m/s = max. velocity (constant)
t seconds = time to reach max. velocity, which is also the time taken in retarding, since the distance travelled during each interval is the same.
s m = distance travelled during accelerating period
$= \frac{1}{4} \times 100$ (quarter of total distance)
$= 25$ m.

$$s = \left(\frac{v + u}{2}\right)t$$

$$25 = \left(\frac{v + 0}{2}\right)t$$

$$\therefore \quad vt = 50 \qquad (1)$$

Time taken to travel at constant velocity v is $(12 - 2t)$ seconds.

Area of trapezium $abcd$ = total distance travelled.

$$\therefore \quad \tfrac{1}{2}[ad + bc] \times ce = 100$$

$$\tfrac{1}{2}[12 + (12 - 2t)] \times v = 100$$

$$12v - vt = 100$$

But $vt = 50$ from equation (1).

$$\therefore \quad 12v - 50 = 100$$

$$v = 12{\cdot}5 \text{ m/s.}$$

Constant velocity during centre portion of lift equals 12·5 m/s.

Example 3.6

The velocity of a body at certain times is given in the accompanying table. Draw the velocity-time graph and find the distance travelled by the body during the first 10 seconds of its motion.

Time, seconds	0	2	4	6	8	10	12
Velocity, m/s	0	2·4	5·6	9·6	14·4	20	26·4

The velocity-time graph is drawn in Fig. 3.9. The distance travelled in 10 seconds is given by the shaded area under the graph up to the ordinate drawn

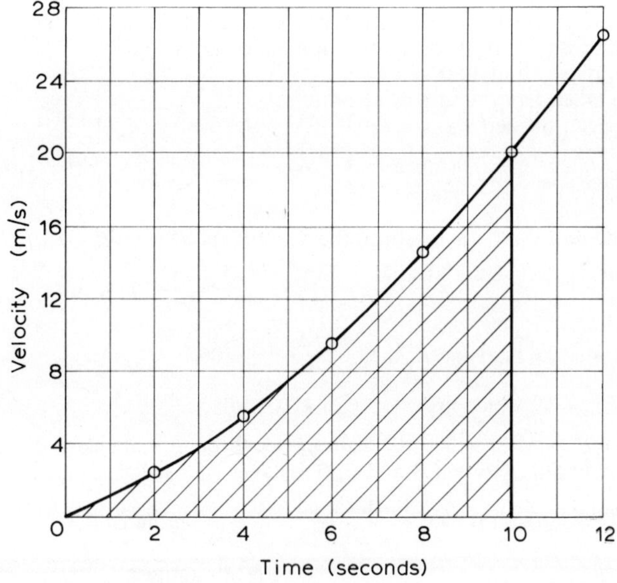

Figure 3.9

through $t = 10$ seconds. This area is obtained by Simpson's Rule (see Appendix 1). The base is divided into 10 equal parts at 1 second intervals.

First and Last	*Odd*	*Even*
0	2·4	1·0
20·0	5·6	4·0
——	9·6	7·6
20·0	14·4	12·0
	——	16·8
	32·0	——
	2	41·4
	——	4
	64·0	——
		165·6

Area $= \frac{1}{3}[20 + 64 + 165{\cdot}6]$
$= 83{\cdot}2$ square units representing a distance of 83·2 m.

The body travels 83·2 m in the first 10 seconds of its motion.

Example 3.7

The driver of an express train is due to arrive at a point A at midday, at a point B, 8 km farther on, at 12.04 p.m., and at C, another 8 km on, at 12.08 p.m. When passing A, he is moving at 120 km/h and is $\frac{1}{2}$ minute ahead of his time. He retards uniformly until reaching point B, when he is exactly on time. If he now maintains this speed, how much out of schedule time will he be when he reaches C?

NOTE. The expression 12.04 p.m. must be taken as 4 minutes past 12 o'clock.
Train passes A at 11.59½ a.m. (½ minute early).
Train passes B at 12.04 p.m.
Time taken from A to B is 4½ minutes.
Distance travelled 8 km.

$$\text{Average speed} = \frac{\text{Distance}}{\text{Time}} = \frac{8 \times 60}{4\frac{1}{2}} = 106{\cdot}7 \text{ km/h.}$$

Since the retardation is uniform, the average speed is given by

$$\frac{\text{Initial} + \text{Final speed.}}{2}$$

If v km/h is the final speed, then $\dfrac{120 + v}{2} = 106{\cdot}7$

which gives $v = 93{\cdot}4$ km/h.

This is the speed at which the train passes through point B, and is the speed which it maintains between B and C, a distance of 8 km.

$$\text{Time taken from B to C} = \frac{\text{distance}}{\text{speed}} = \frac{8 \times 60}{93{\cdot}4} \text{ minutes}$$

$$= 5{\cdot}14 \text{ minutes.}$$

∴ Train reaches C at 12 hours 4 minutes + 5·14 minutes,
i.e. 12 hours 9·14 minutes

which is 1·14 minutes after schedule.

EXERCISE 3

1 At a given instant a car is moving with a velocity of 5 m/s and a uniform acceleration of 2 m/s². What will be its velocity after 10 seconds?

2 A car passes a certain post with a velocity of 5 m/s and another post 1 km farther on with a velocity of 15 m/s, the acceleration being uniform. What is the average velocity of the car? How long will it take the car to travel the distance between the posts? What will be the acceleration of the car?

3 Two signals are 850 m apart. A train passes the first with a speed of 100 kilometres per hour and, applying the brakes to give a uniform retardation, passes the second signal 40 seconds later. Determine the velocity of the train passing the second signal, and also the magnitude of the retardation.

4 With what initial velocity must a body be travelling in order to come to rest in 30 m with a uniform retardation of 2·5 m/s²?

5 A bullet travelling with a velocity of 300 m/s strikes a target and penetrates 12 cm. Assuming the retardation is constant, determine its value and also the time taken for the target to stop the bullet.

6 The maximum retardation with which the brakes of a train can reduce its speed is 1·8 m/s². If the train is travelling at 100 km/h, what is the shortest distance, and time, in which the train can be brought to rest?

7 The driver of a train is due to arrive at a point A at 1 p.m., at a point B, 15 km farther on, at 1.10 p.m., and at C, another 15 km on, at 1.20 p.m. When passing A he is moving at 80 km/h and is two minutes behind time.

The train is accelerated uniformly, so that he passes C exactly on time. How much out of schedule time will he be when passing B?

8 A car starts from rest, and travels a distance of 5 km in 5 minutes. During the first part of the journey the car accelerates at 1·5 m/s^2 and reaches a maximum velocity at which it travels during the second part of the journey. The car then retards at 2·5 m/s^2, and is brought to rest at the end of the 5 km. Determine the constant velocity during the second portion of the journey.

9 The maximum acceleration and retardation of a lift is 1·5 m/s^2. The maximum velocity is 3·5 m/s. What is the least time from rest to rest to ascend 30 m.

10 A lift rises 30 m in 9 seconds. For the first quarter of the distance the lift is uniformly accelerated, and during the last quarter it is uniformly retarded. If the speed is constant during the centre portion of the ascent, determine its value. [N.C.T.E.C.]

11 The velocity of a body at certain times is given in the accompanying table. Draw the velocity-time graph and find the distance travelled by the body during the first 5 seconds of its motion.

Time, seconds	0	1	2	3	4	5	6
Velocity, m/s	0·0	4·5	9·0	13·5	18·0	22·5	27·0

12 A body having an initial speed of 5 m/s and moving with a uniform acceleration covers a distance of 560 m in 60 seconds. Calculate the acceleration and the maximum speed. [N.C.T.E.C.]

13 A mine cage starts from rest at the bottom of the shaft and the speed rises uniformly in the following 12 seconds until a value of 5 m/s is reached. This speed is maintained until a height of 200 m above the bottom is reached. Find the distance covered during the initial period of 12 seconds when the speed is increasing, and also the time taken to reach 200 m from rest. [N.C.T.E.C.]

14 A boy on a bicycle stops pedalling and the machine is uniformly retarded. In half a minute he covers 100 m and comes to rest in a further 30 seconds. Sketch the speed-time graph and determine (*a*) the speed of the machine when he stopped pedalling, (*b*) the total distance covered after stopping pedalling. [N.C.T.E.C.]

15 A train starting from rest reaches a speed of 70 km/h after travelling 5 km. Assuming uniform acceleration, calculate its value. [N.C.T.E.C.]

16 Three telegraph posts, A, B and C, stand by the side of a road, the distances AB and BC being each 50 m. A motor-car proceeding with uniform acceleration passes post A and then takes 8 seconds to reach post B and a further 7 seconds after passing post B to reach post C. Calculate (*a*) the acceleration of the car in m/s^2, (*b*) the speed of the car at A and B. [N.C.T.E.C.]

17 Define 'average velocity'. A motor-car, which is travelling with uniform acceleration, is observed to cover distances of 6 m and 7·5 m respectively in two consecutive intervals of 1 second each. Find the value of the uniform acceleration. [U.L.C.I.]

18 A car moving with uniform acceleration passes three posts spaced at 150 m intervals. The time to travel between the first two posts is 8 seconds and that between the second two posts is 6 seconds. Calculate the acceleration of the car and the speed when passing each post. [N.C.T.E.C.]

19 A mine cage starting from rest at the bottom of the shaft increases its speed

uniformly for the first 15 seconds until a speed of 8 m/s is reached. The speed is then maintained constant for a period of 5 seconds and finally the cage is brought to rest at the top of the shaft after a further period of 10 seconds uniform retardation.
Determine: (*a*) the initial acceleration, (*b*) the final retardation, (*c*) the depth of the shaft.
Draw a speed-time graph of the motion. [N.C.T.E.C.]

Chapter Four
Relative Velocity

Let us consider three of the ways in which we can observe the motion of a train B travelling at 60 km/h. In each case we will make our observations as we are seated in the coach of a second train A.

1 Our train A is at rest in the station. Then, on a parallel track, comes train B speeding past the windows, and we have the experience of seeing a train moving at 60 km/h.
2 Our train is moving at 60 km/h and train B, on a parallel track, is coming towards us at 60 km/h. As it flashes past the window we have the experience of watching a train apparently moving at 120 km/h. It is a remarkable fact that trains which meet us always appear to be travelling faster than we do.
3 Our train is still moving at 60 km/h and train B, still on the parallel track, is now travelling in the same direction as we are. For quite a while now the door of a compartment on the second train has remained opposite to the door in our compartment. In fact, the second train does not appear to be moving at all. It seems that it would be so easy to step from one train to the other.

The velocity of the train B, although being the same in each case, does in fact appear to vary, and we only see the real velocity when we make our observation from a stationary position. The velocity which the train appears to have is always dependent upon the velocity of the observation point. We say that a body has a certain velocity relative to another body, and by that we mean that the first body appears to have such a velocity when observed from the second body.

Or if we think of two bodies A and B each having independent motion, then the velocity of B relative to A is the velocity which B *appears* to have to an observer travelling on A.

Let A and B be two bodies whose velocities v_a and v_b are in the directions shown in Fig. 4.1. We require to find the velocity of B relative to A, i.e. the velocity with which B appears to be moving when observed from the moving point A. Notice that the required velocity is to be relative to A. We must consider a method of bringing A to rest. Imagine that A and B are moving over a sheet of paper. If the paper is moved with a velocity equal and opposite to that of A (i.e. with a velocity of $-v_a$) the effect will be that A is brought to rest. As fast as A moves forward, the paper moves backward with the same velocity, and consequently A makes no progress, at any rate as far as the earth is concerned. You will see, however, that A would ultimately reach the end of the paper. The same effect would be produced if you were running forward on an endless belt moving with a velocity equal but opposite to that of your velocity. As far as the surrounding structure was concerned you would be stationary, moving neither forwards nor backwards. The effect as far as the belt was concerned would be that you were running forward with a certain velocity.

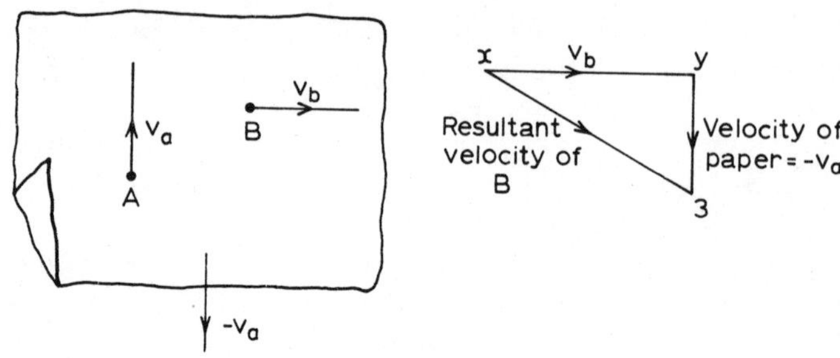

Figure 4.1 *Relative Velocity*
The paper is moved so that A is brought to rest. This requires that the paper moves with a velocity equal and opposite to that of A. This affects the velocity of B.

Now consider what is happening to B in Fig. 4.1. Not only is it moving with its original velocity v_b but it has also added to it the velocity of the paper, which is $-v_a$. The resultant velocity of B can be obtained from the velocity triangle.

xy represents the velocity of B, v_b;
yz represents the velocity of the paper which is equal and opposite to that of A so that $yz = -v_a$;
xz represents the resultant velocity of B.

We now have A brought to rest and B moving with a velocity represented by xz. So that if you were situated on A, then the velocity which B would appear to have would be xz.

Hence xz is the velocity of B relative to A.

Referring again to the velocity triangle, we have

$$xz = xy + yz \quad \text{or}$$

Velocity of B relative to A = velocity of B + velocity of the paper
= velocity of B + (−velocity of A)
= velocity of B − velocity of A
= difference of the velocities.

DEFINITION. **The velocity of B relative to A is the vectorial difference of the two velocities.**

This can be written in symbols as follows, provided that we take v_{ba} to mean the velocity of B relative to A:

$$v_{ba} = v_b - v_a$$

It may be useful to note at this point that:

The resultant of two velocities is the vectorial sum of the velocities, while
The relative velocity is the vectorial difference of the velocities.

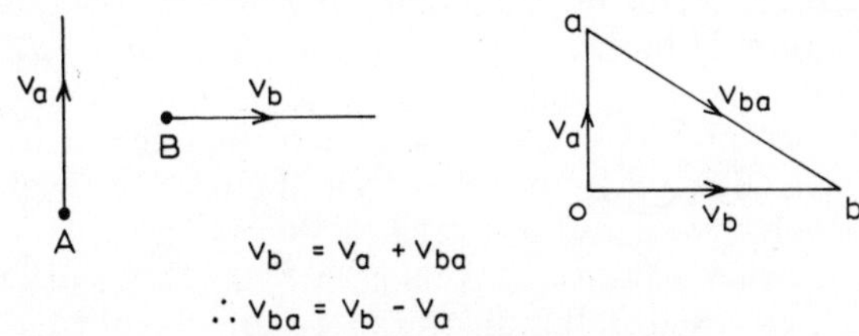

Figure 4.2 *Relative Velocity*
The velocity of B relative to A is the velocity of B minus the velocity of A.

The basic method of drawing the velocity triangles is shown in Fig. 4.2.

oa represents the velocity of A,
ob represents the velocity of B,
ba represents the velocity of A relative to B,
ab represents the velocity of B relative to A.

Note that the letter to which the arrow points indicates the velocity under consideration. In the vector ***ab*** the arrow points to ***b***, hence the vector represents the velocity of B relative to A.

All velocities drawn from *o* are absolute velocities (their definition is given in the next section), while all vectors drawn from points other than *o* are relative velocities.

Now let us apply this general rule to the matter of the two trains which began this discussion.

Case 1. Velocity of our train A = 0 km/h
Velocity of train B = 60 km/h
Velocity of train B relative to A = 60 − 0
= 60 km/h.

Case 2. Velocity of our train A = +60 km/h
Velocity of train B = −60 km/h (opposite direction)
Velocity of train B relative to us = −60 − (+60)
= −120 km/h.
(the negative sign implies that the direction is opposite to that of our train).

Case 3. Velocity of our train A = +60 km/h
Velocity of train B = +60 km/h (same direction)
Velocity of train B relative to A = +60 − (+60)
= 0 km/h.

NOTE. You will have to make sure that you subtract A from B.

Velocity of B relative to A = velocity of B − velocity of A

The order of the letters B . . . A is the same on both sides of the equation.

Absolute velocity

By absolute velocity we mean the real or true velocity of the body as observed from the earth. Fundamentally it is true to say that due to the compound nature of the earth's movement even a body at rest on the earth is actually moving in space. But for our considerations we shall assume that the earth is a fixed datum and that the velocity which a body has when observed from the earth is the true or absolute velocity. It is in fact a velocity relative to the earth.

When we say that a train is travelling at 60 km/h we mean that when observed from the fixed station platform the train is moving at 60 km/h, and this is its absolute velocity. If you walk along the corridor at 5 km/h, which really is your velocity relative to the train, then your absolute velocity or your velocity with respect to a fixed point on the earth is either $60 + 5 = 65$ km/h or $60 - 5 = 55$ km/h, dependent upon whether you are moving to the front or the back of the train. You would certainly be more conscious of your real absolute velocity if you were to walk along the roof of the carriage rather than inside the corridor.

Example 4.1

Determine the velocity of a ship B relative to a ship A if B is travelling at 10 km/h in direction 30° east of north, and A is travelling at 15 km/h 60° east of south.

METHOD 1

Fig. 4.3(i) shows the velocities of the two ships. Since we require to know B's velocity relative to A, we must imagine A brought to rest by thinking of the sea moving with the same velocity as A but in the opposite direction (i.e. 15 km/h in direction west of north). The effect of this is to give B two velocities:

(*a*) 10 km/h 30° east of north, and
(*b*) 15 km/h 60° west of north.

These velocities are represented respectively by xy and yz in Fig. 4.3(ii).

The resultant, xz, representing 18·03 km/h 26° 18′ west of north is the velocity with which B appears to be moving when viewed from the now stationary ship A, i.e. the velocity of B relative to A.

METHOD 2—*Use of Basic Equation*

Velocity of B relative to A = velocity of B − velocity of A
(Note the order of the letters B, A.)
= velocity of B + (− velocity of A)
= velocity of B + reverse velocity of A

1 Draw xy to represent velocity of B to scale: 10 km/h 30° east of north
2 Draw yz to represent reverse velocity of A to scale: 15 km/h 60° west of north.

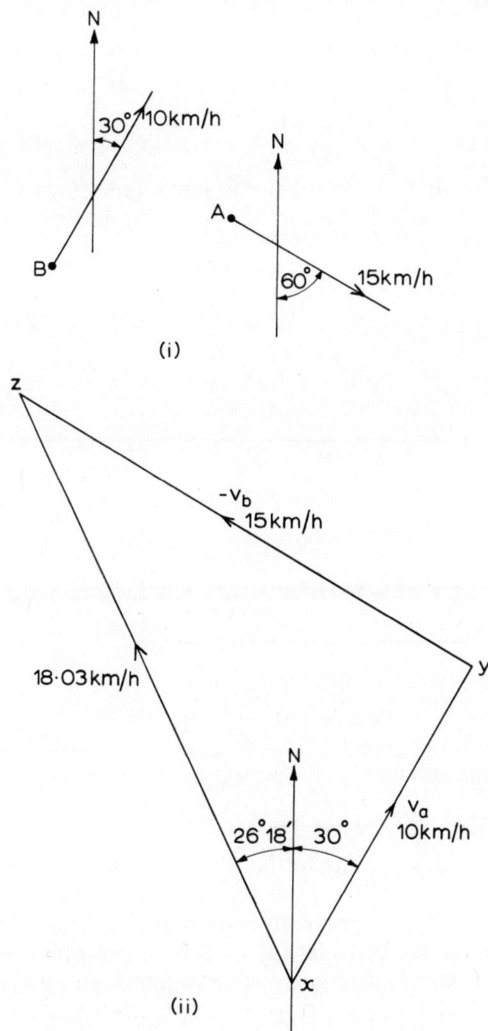

Figure 4.3

3 The velocity of B relative to A = ***xz***
= 18·03 km/h at 26° 18′ east of north.

Notice particularly the direction of the vector ***xz***.
xz is the sum of ***xy*** and ***yz***, and is drawn from the start *x* to the finish *z*.

Special case of relative velocity

The basic equation for examples on relative velocity is

Velocity of A relative to B = velocity of A − velocity of B

This can be rewritten in the form:

Velocity of B = velocity of A + velocity of B relative to A

We can use this form, if we require to know the velocity of any point B, if we are given the velocity of another point A and we know the velocity of B relative to A.

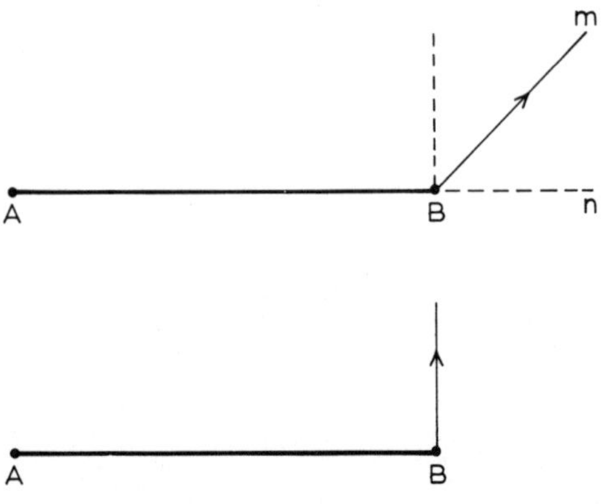

Figure 4.4 *Motion of Rigid Rod*

Any motion which B has relative to A must be in a direct perpendicular to the length of the rod; otherwise the rod will either stretch or collapse.

In the case of ships moving in the water and aeroplanes moving in the air, the magnitude and direction of the velocity of one relative to another can be any value, since there is no restriction on the movement of these vessels. If, however, we consider the motion of two points on a rigid link, such as a crank or a connecting-rod of an engine, then we shall find that the velocity of these two points is very restricted, at least as far as direction is concerned. AB is a rigid rod, Fig. 4.4. This means that the distance AB is fixed. It cannot be increased as though it were made of elastic, neither can it be decreased as if it were a telescope.

Now suppose that the velocity of B relative to A is along the line B*m*, the direction of which was made at random. The component of this velocity along the length of the rod is B*n*, which means that B is moving away from A with a velocity B*n*, i.e. the length of AB is increasing. But since the rod is rigid, AB cannot increase. There cannot therefore be a *Bn* component. You can apply the same argument if the direction of B*m* were such as to make the *Bn* component be directed towards A. If there is no B*n* component, then the direction of the relative velocity *Bm* must be perpendicular to AB. This leads to a very important rule.

In any rigid body, the velocity of one point in the body relative to another point in the body is always in a direction perpendicular to the line joining the points.

This rule will apply in all cases, irrespective of the type of motion which the body may be making. For example, consider the crank and piston mechanism of an engine (Fig. 4.5). The crank OC rotates with a uniform

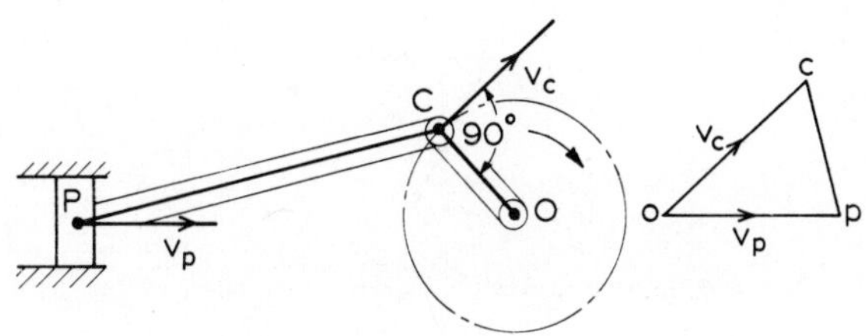

Figure 4.5 *Velocity Diagram for Engine Mechanism*

velocity about O. The piston slides backwards and forwards along the horizontal line PO, being driven by the connecting rod CP. One end, C, of the connecting-rod moves in a circular path, and the other end, P, moves in a straight line. You may have carried out an exercise in your drawing class to show that the motion of any other point on the connecting-rod is on an elliptical path. Well, the velocity of C will always be tangential to the path on which it moves, i.e. it will always be at right angles to OC. Knowing the angular velocity and the length of the crank OC, we shall be able to determine the magnitude of v_c.

Now, velocity of P = velocity of C + velocity of P relative to C.

Draw ***oc*** to represent the magnitude and direction of C's velocity.

Draw ***op*** to represent the direction of P's velocity. Since the magnitude is not known, we cannot fix a length to ***op***.

Through *c* draw a line to represent the velocity of P relative to C, its direction being perpendicular to PC. Again the length of ***pc*** cannot be fixed.

The intersection of these lines gives the point *p* such that ***op*** is equal to the velocity of the piston P.

Rotation of link

Fig. 4.6 shows a link AB moving in one plane so that A has a velocity v_a and B has a velocity v_b. Then the velocity of B relative to A is ***ab***. If we can imagine that we are standing on point A and looking towards B, we should see that B moved in the direction of ***ab***, a direction which has been shown dotted on the end of the link. The direction is down and to the right. This means that the rod AB is turning in a clockwise direction. If you were to be standing at B, the point A would appear to be moving in direction ***ba***, which is in a directly opposite sense to the one just mentioned. But you will see that the direction of rotation is still the same—clockwise.

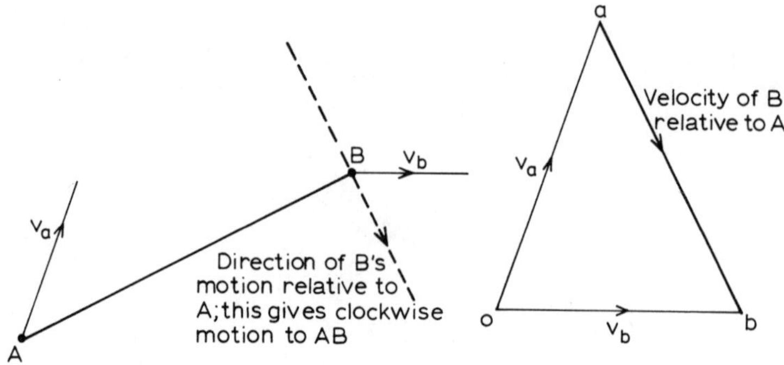

Figure 4.6 *Velocity Diagram*

Angular velocity of a link

The angular velocity of any link rotating about a fixed centre is equal to

$$\frac{\text{linear velocity of end of link}}{\text{length of link}}.$$

The angular velocity of a link which is not rotating about a *fixed* axis is equal to

$$\frac{\text{velocity of one end relative to the other}}{\text{length of link}}.$$

In the case of the link AB in Fig. 4.6

$$\text{Angular velocity of AB} = \frac{v_{ab}}{\text{AB}} = \frac{ab}{\text{AB}}$$

In addition to the angular velocity, the link also has a linear velocity. It is usual to indicate this linear velocity by stating the linear velocity of the centre of gravity of the link.

Velocity of point on link

Sometimes we require to know the velocity of a point on a link other than the two end points. This can easily be obtained from the velocity diagram. In Fig. 4.7, *aob* is the velocity triangle for the link AB with ***oa*** and ***ob*** representing the velocities of A and B respectively and ***ab*** having been drawn perpendicular to AB representing the velocity of B relative to A. We require to know the velocity of a given point C on the link. On the line ab, determine the point c such that $\frac{ac}{ab} = \frac{AC}{AB}$. Join oc, then $\boldsymbol{oc}$ represents in magnitude and direction the velocity of C.

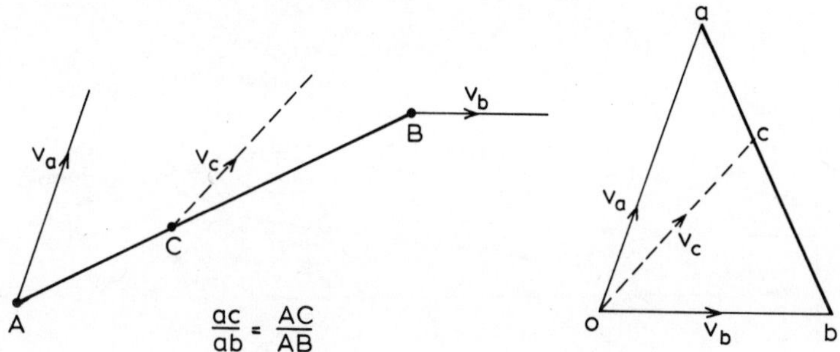

Figure 4.7 *Velocity of Point on Link*

We know that $v_c = v_a + v_{ac}$. The velocity of C relative to A will be perpendicular to CA, and will be proportional to the distance from C to A. Thus, if C were half-way between A and B, then the velocity of C relative to A would be half that of B relative to A. Hence by obtaining the point c such that $\frac{ac}{ab} = \frac{AC}{AB}$ the vector $\boldsymbol{ac}$ satisfies both conditions of direction and magnitude, and thus represents the velocity of C relative to A. Therefore $\boldsymbol{oc}$, which is the vector sum of $\boldsymbol{oa}$ and $\boldsymbol{ac}$, will represent the velocity of C.

Example 4.2

The crank and connecting-rod of an engine are 100 mm and 500 mm long respectively. The crank rotates at 300 rev/min. Determine (*a*) velocity of piston, (*b*) velocity of point on centre line of connecting-rod and 100 mm from crank-pin centre, and (*c*) angular velocity of the connecting-rod for the positions in which the crank has turned through 45° and 135° from inner dead centre, i.e. from the position when the piston is farthest from the crankshaft.

The solution applies to both diagrams, the upper one showing 45° angle and the lower one showing 135° angle (Fig. 4.8).

$$\text{Angular velocity of crank} = \omega = \frac{300}{60} \times 2\pi$$
$$= 31{\cdot}42 \text{ rad/s}$$

$$\text{Velocity of C} = \omega \times \text{OC}$$
$$= 31{\cdot}42 \times 0{\cdot}1$$
$$= 3{\cdot}142 \text{ m/s.}$$

1 *Velocity of* C. Magnitude, 3·142 m/s; direction, perpendicular to OC. Draw vector $\boldsymbol{oc}$ to represent this.

2 *Velocity of* P *relative to* C. Magnitude, not known; direction, perpendicular to link CP. This is a relative velocity, therefore the vector is drawn through a

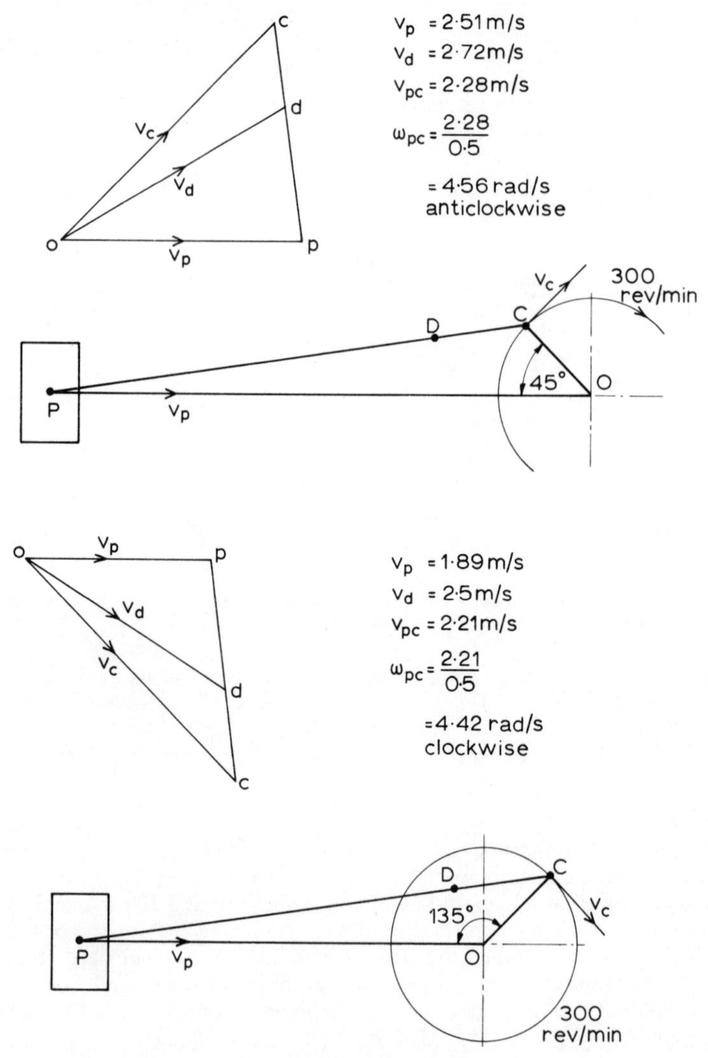

Figure 4.8

point other than *o*. Draw a line through *c* perpendicular to CP to represent the direction of this velocity.

3 *Velocity of* P. Magnitude, not known; direction, since P moves in a straight line, the direction of its velocity is in the direction of its displacement. This is an absolute velocity, therefore the vector is drawn through the point *o*. Through *o* draw a vector parallel to P's motion. This vector will intersect the vector drawn in 2 in the point *p*. Then ***op*** represents in magnitude and direction the velocity of the point *p*.

4 *Velocity of* D *distance* 0·1 m *from* C. Obtain the point *d* on the line *cp* so that

$\frac{cd}{cp} = \frac{CD}{CP}$. Join *od*. Then ***od*** represents in magnitude and direction the velocity of the point D.

5 *Angular velocity of the connecting-rod.* This is equal to $\frac{v_{pc}}{CP} = \frac{cp}{CP}$.

In the case of the 45° angle $v_{pc} = 2{\cdot}28$ m/s

Hence $$\omega = \frac{v_{pc}}{CP} = \frac{2{\cdot}28}{0{\cdot}5} = 4{\cdot}56 \text{ rad/s}$$

Note that the direction of P's velocity relative to C is given by the vector ***cp*** which is down and slightly to the right. If you imagine P moving in this direction when viewed from C, you will see that the direction of rotation is anti-clockwise.

In the case of the 135° angle $v_{pc} = 2{\cdot}21$ m/s

Hence $$\omega = \frac{v_{pc}}{CP} = \frac{2{\cdot}21}{0{\cdot}5} = 4{\cdot}42 \text{ rad/s}$$

In this case the direction of P's velocity relative to C is given by the vector ***cp*** which is up and slightly to the left. If you imagine P moving in this direction when viewed from C you will see that the direction of rotation is clockwise.

EXERCISE 4

1 Determine the velocity of a ship A relative to a ship B if A is travelling at 15 km/h in a direction north-east and B is travelling at 20 km/h in a direction due south.

2 Fig. 4.9 indicates a rigid link AB 3 m long in which end A is travelling at 3 m/s due north, whilst B is moving on a line in the east–west direction. Determine the velocity of B and also the angular velocity of the rod AB.

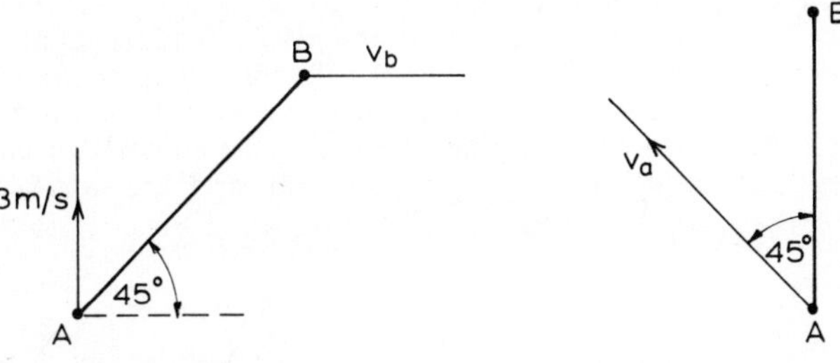

Figure 4.9 **Figure 4.10**

3 The link AB in Fig. 4.10 has an angular velocity of 10 rad/s clockwise. If the link is 5 m long and the velocity of A is 4 m/s in the direction shown, determine the magnitude and direction of the velocity of B.

4 The velocity of the point B (Fig. 4.11) on the rigid link ABC is 5 m/s in the direction shown. If AB is 4 m and BC is 3 m, determine the velocity of A, whose direction is indicated, and of C, and also the angular velocity of the rod.

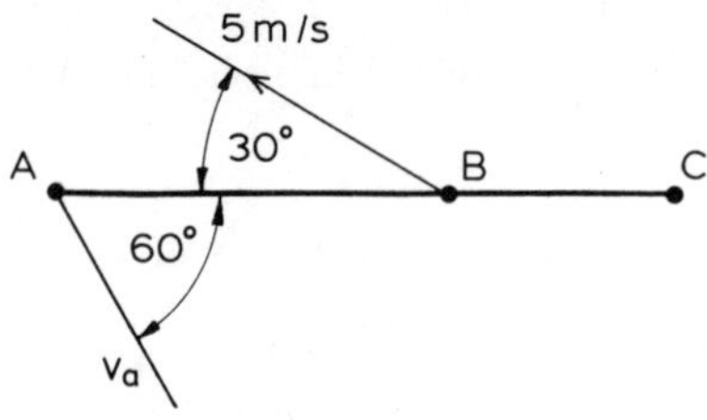

Figure 4.11

5 Find the velocity of a ship A relative to a ship B if A is travelling 20° east of south at 12 km/h and B is travelling 60° east of north at 20 km/h. [N.C.T.E.C.]

6 Explain what is meant by relative velocity. A ship A is travelling due south at 15 km/h and another, B, due south-west at 24 km/h. Find the velocity of B relative to A in magnitude and direction [N.C.T.E.C.]

7 The crank and connecting-rod of an internal-combustion engine are 60 mm and 270 mm long respectively. The crank rotates clockwise at 450 rev/min. When the crank has turned through 45° from the inner dead centre, determine: (*a*) the velocity of the piston, (*b*) the magnitude and direction of the velocity of a point on the centre line of the connecting-rod and 120 mm from the crank-pin centre, (*c*) the angular velocity of the connecting-rod. [U.L.C.I.]

8 The crank and connecting-rod of an internal-combustion engine are 20 mm and 80 mm long respectively. The crank rotates at 1200 rev/min. When the crank has turned through 40° from the inner dead centre, determine: (*a*) the velocity of the piston, (*b*) the velocity of a point on the centre-line of the connecting-rod and 30 mm from the crank-pin centre, and (*c*) the angular velocity of the connecting-rod. [U.L.C.I.]

9 The connecting-rod of a reciprocating engine is 1·2 m and the crank 300 mm. When the crank is in the position 150° past the inner dead centre the piston speed is 4·5 m/s. Determine: (*a*) the speed of the crankshaft in revolutions per minute, (*b*) for the crank position stated, (i) the angular velocity of the connecting-rod, and (ii) the velocity, in magnitude and direction of the mid-point along the axis of the connecting-rod. [U.L.C.I.]

Chapter Five
Motion of Falling Bodies and Projectiles

Motion of falling bodies

In or about the year A.D. 1590 Galileo carried out a rather surprising experiment. At the same instant he dropped two masses, a 1 kg and a 50 kg, from a point 30 metres above the ground level. Actually he made use of the famous leaning tower of Pisa in Italy. Now for two thousand years people who gave any thought to the matter believed that heavier masses fell faster than lighter masses. They believed, for example, that the 50 kg mass would travel through 50 metres while the 1 kg mass was falling through 1 metre. The observers of the experiment were therefore very much surprised to find that the two masses hit the ground practically at the same instant. By careful observation, they did in fact notice that the 50 kg did arrive about 30 mm in front of the 1 kg mass.

Figure 5.1 *Neck and Neck at Pisa*

It was generally agreed amongst the ancients that a heavy body would fall to the earth more quickly than a lighter body. Galileo surprised a lot of people when he dropped two different masses from the Leaning Tower of Pisa. The two masses reached the ground practically simultaneously, causing people to completely revolutionise their ideas on mechanics.

The result of this experiment together with those of similar experiments carried out in this country by Sir Isaac Newton, a hundred years later, led scientists to the conclusion that when a body is allowed to fall freely to the ground, the acceleration of the body is approximately 9·82 m/s^2. *Whatever the size, whatever the shape, provided that the effects of wind resistance are ignored, the acceleration is the same.* The magnitude of the acceleration does, however, vary at different points on the surface of the earth. We usually denote this particular acceleration by the letter g.

In London the value of g is 9·81 m/s^2.

At the Equator the value of g is 9·78 m/s^2.
At the North or South Pole g is 9·83 m/s^2.

Positive and negative velocities and accelerations

If we throw a stone into the air with an initial velocity of v m/s, we know that the velocity will decrease until the stone reaches a maximum height, when it will begin to fall again with an increasing velocity. The direction of the velocity will be opposite when falling to what it was when rising. In order to allow for this, we shall have to take into account the sign of the velocity by choosing one direction to be the positive direction.

It will be most convenient to choose the upward direction as the positive direction.

Therefore, upward velocities and upward distances are all positive. Downward velocities and downward distances are all negative.

Since the acceleration due to gravity is always acting towards the centre of the earth, i.e. as far as we are concerned in the downward direction, it will always be considered to be negative. With this in mind we can use the same expressions which we developed on page 14 by inserting $-g$ for f.

Formula for falling bodies

$$v = u - gt$$
$$v^2 - u^2 = -2gs$$
$$s = ut - \tfrac{1}{2}gt^2.$$

Upward velocities and distances are *positive*.
Downward velocities and distances are *negative*.

Note regarding the value of s

The value of s in these expressions is always the distance of the body at a given instant, measured from the starting-point. This is not necessarily the same as the distance through which the body has travelled.

For example, if a stone is thrown to a height of 100 m and at the given instant has fallen back through 80 m then the actual distance through which the stone has travelled is 180 m. But the distance of the stone from the starting-point at that instant is 20 m. This will be the value of s as given in the above formula.

If the stone is thrown upwards from the top of a tower or cliff so that it can fall to a level below that of the starting-point, then the value of s will increase to a maximum whilst the stone is moving to the top of its flight; it will then decrease to zero as the stone returns to its initial level. The value of s will then become negative as t further increases, indicating that the stone is now below the level from which it was thrown.

In all cases s gives the distance measured from the starting-point to the position of the stone at any value of t. See Example 5.4, page 42.

Graphical representation of velocity and acceleration

Velocity and acceleration are both vector quantities, i.e. they involve both magnitude and direction. Hence they can be represented by a directed line in a similar manner to that of representing forces. They are likewise subject to the same methods of treatment as forces, particularly with regard to addition, subtraction and resolution into components.

For example, if you walk across the deck of a ship with a velocity of 3 m/s while the ship is moving with a velocity of 4 m/s, your actual velocity is the sum of two velocities. This sum must be obtained vectorially, i.e. by taking direction into account. Draw the vector ***oa*** to represent the ship's velocity, 4 m/s. Draw the vector ***ab*** to represent your velocity across the deck of the ship, 3 m/s, then your actual velocity is given by the vector ***ob***, which represents to scale 5 m/s inclined at approximately 37° to the direction of the ship (Fig. 5.2).

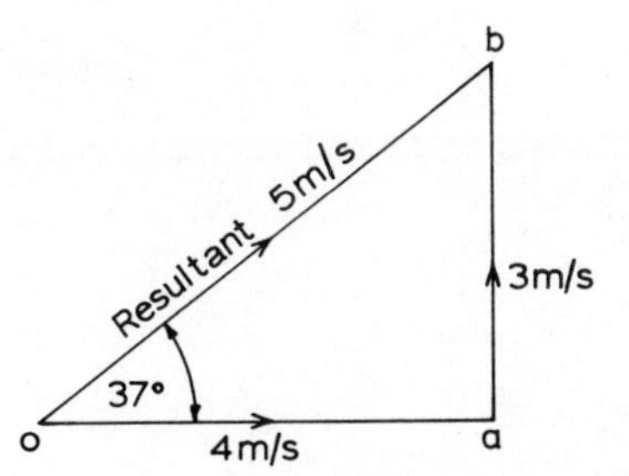

Figure 5.2 *Resultant Velocity*

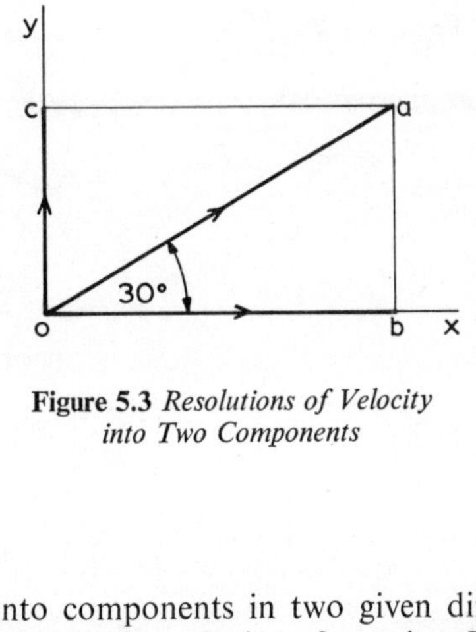

Figure 5.3 *Resolutions of Velocity into Two Components*

Components of velocity

Again, we can divide a velocity into components in two given directions. The vector ***oa*** in Fig. 5.3 represents the velocity of a point. It is required to know its component velocities in two directions at right angles to each other, one of which is at 30° to the ***oa*** direction. Through *o* draw the line *ox* at 30° to *oa*, and the line *oy* at 90° to *ox*. Through *a* draw *ab* parallel to *oy*, and *ac* parallel to *ox*. Then ***ob*** is the component in the *ox* direction, and ***oc*** is the component in the *oy* direction. It will be clear that ***ob*** = ***oa*** cos 30° and ***oc*** = ***oa*** sin 30°. Since most problems on components deal with directions at right angles to each other, this trigonometrical relationship will always hold (Fig. 5.4).

$$\text{Horizontal component} = \text{velocity} \times \cos\theta$$
$$\text{Vertical component} = \text{velocity} \times \sin\theta$$

where θ is the angle which the velocity makes with the horizontal direction.

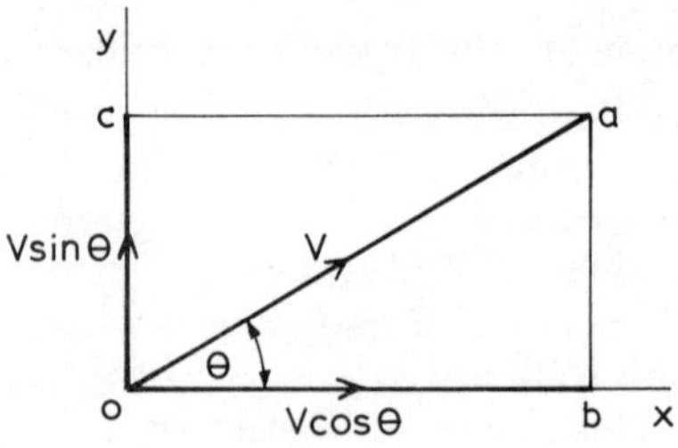

Figure 5.4 *Resolution of Velocity into Two Components*

Example 5.1

An aircraft is climbing at a velocity of 800 km/h inclined at 30° to the horizontal. What is its forward velocity? How high will it climb in the next 30 seconds assuming the velocity remains constant?

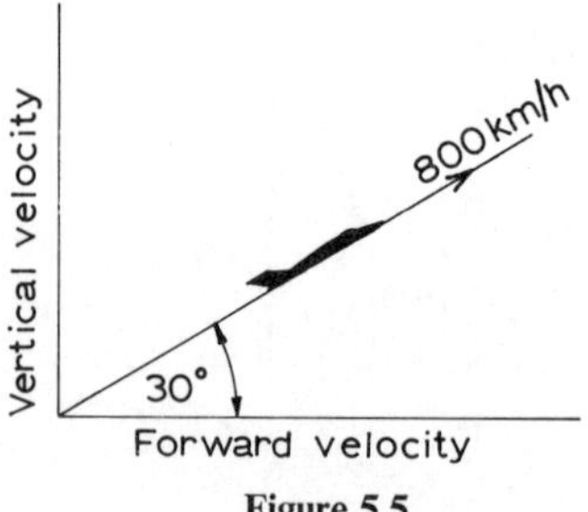

Figure 5.5

$$\text{Forward velocity} = 800 \cos 30° = 692{\cdot}8 \text{ km/h.}$$

$$\text{Vertical velocity} = 800 \sin 30° = 400 \text{ km/h.}$$

$$\text{Vertical distance travelled in 30 seconds} = \frac{400 \times 30}{60 \times 60} = 3{\cdot}33 \text{ kilometres.}$$

Forward velocity of aircraft is 400 km/h; it will move through a vertical distance of 3·33 km in next 30 seconds.

Projectiles

If an object is thrown or projected from a cliff 125 m high in a horizontal direction, with a velocity of 50 m/s, its subsequent motion is made up of two parts.

(1) It will always have a forward velocity of 50 m/s. The horizontal distance travelled will be dependent upon this horizontal velocity and the time elapsed from the start.

(2) Coming under the influence of gravity immediately it leaves the cliff.

it will have a vertical velocity which will be zero at the instant of projection, but which will increase by 9·81 m/s for every second of motion. In other words, its vertical motion is identical to that of a falling body.

Consequently, the actual velocity of the object will be the vector sum of these horizontal and vertical components. The horizontal velocity will always be 50 m/s. The vertical velocity will be given by

$$v = 0 - 9{\cdot}81t$$
$$v = -9{\cdot}81t \text{ m/s}$$

where t is the time from the instant of projection.

The path of the object can be plotted by calculating the horizontal distance travelled given by $x = 50t$ and the vertical distance travelled given by $s = ut - \frac{1}{2}gt^2$, and since there is no initial vertical velocity, this expression becomes $s = -\frac{1}{2}gt^2$, where g can be taken as 9·81 m/s^2.

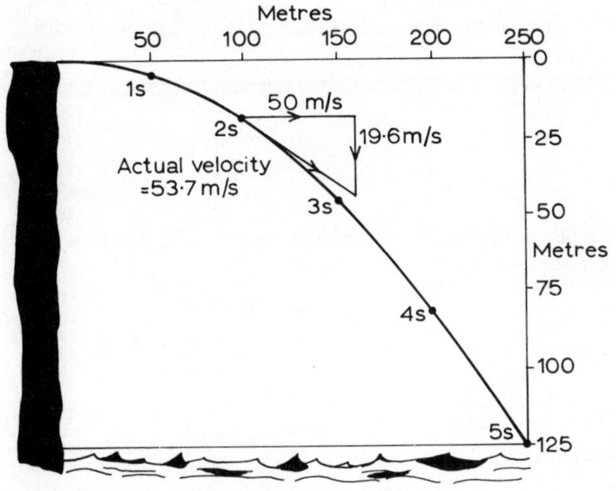

Figure 5.6 *Path of Projectile*

Horizontal velocity is constant. Vertical velocity increases in the same way as that of a falling body dropped from rest.

time	0	1	2	3	4	5 s
horizontal distance	0	50	100	150	200	250 m
vertical distance	0	4·91	19·6	44·1	78·5	122·6 m
horizontal velocity	50	50	50	50	50	50 m/s
vertical velocity	0	9·81	19·6	29·4	39·2	49·1 m/s
actual velocity	50	50·95	53·7	57·98	63·54	70·7 m/s

The actual velocity v is given by $v = \sqrt{v_h^2 + v_v^2}$, which is an application of the theorem of Pythagoras to the vector triangle. It will be seen from Fig. 5.6 that the path covered by the object is a parabola which is characteristic of paths covered by projectiles. At a later stage we shall review the method of dealing with motion of projectiles whose original velocity was not horizontal. For example, the motion of a shell fired into the air at a given inclination to the horizontal.

In the meantime the important things to keep in mind when dealing with projectiles having an original horizontal velocity are:

(1) horizontal motion is determined by initial horizontal velocity, and
(2) vertical motion is identical to that of a falling body dropped from rest.

Example 5.2

A stone is allowed to fall from the edge of a cliff 400 m high. What will be its velocity after 2 seconds?

How far will it have travelled at the end of 4 seconds?
How long will it take the stone to reach the base of the cliff?
What will be the striking velocity on reaching the base of the cliff?

Since the stone is just allowed to fall, it has no initial velocity.
$\therefore \quad u = 0.$

Velocity after 2 seconds:

$$\begin{aligned} v &= u - gt \\ &= 0 - 9{\cdot}81 \times 2 \\ &= -19{\cdot}62 \text{ m/s} \end{aligned}$$

i.e. its velocity is 19·62 m/s downwards.

Distance after 4 seconds:

$$\begin{aligned} s &= ut - \tfrac{1}{2}gt^2 \\ &= (0 \times 4) - (\tfrac{1}{2} \times 9{\cdot}81 \times 4^2) \\ &= -\ 78{\cdot}48 \text{ m} \end{aligned}$$

i.e. the stone has travelled 78·48 m downwards from the point at which it was released.

Time taken to reach base:

In this case

$$\begin{aligned} u &= 0 \text{ m/s} \\ s &= -400 \text{ m, i.e. 400 m down from point of projection.} \\ s &= ut - \tfrac{1}{2}gt^2 \\ -400 &= 0 - \tfrac{1}{2} \times 9{\cdot}81 \times t^2 \\ t &= 9{\cdot}04 \text{ seconds.} \end{aligned}$$

Striking velocity:

$$\begin{aligned} u &= 0 \text{ m/s} \\ s &= 400 \text{ m} \\ v^2 - u^2 &= -2gs \\ v^2 - 0 &= -2 \times 9{\cdot}81 \times -400 \\ v &= 88{\cdot}6 \text{ m/s.} \end{aligned}$$

After 2 seconds the velocity of the stone will be 19·62 m/s.
It will have travelled 78·48 m in the first 4 seconds.

It will reach the base of the cliff in 9·04 seconds, when its striking velocity will be 88·6 m/s.

Example 5.3

A body is projected upwards with a vertical velocity of 120 m/s. Find:
(*a*) when its velocity is 60 m/s,
(*b*) the time taken to reach the maximum height,
(*c*) the maximum height reached, and
(*d*) the time of flight, or the time to reach the ground.

The initial velocity u is + 120 m/s, since its direction is upwards.

(*a*) When velocity is 60 m/s.

There will be two answers to this part, since the body will have a velocity of +60 m/s on its upward flight and −60 m/s on the return flight.

(i) When velocity is 60 m/s moving upward,

$$\begin{aligned} \text{i.e.}\quad v &= +60 \text{ m/s} \\ u &= +120 \text{ m/s} \\ t &= ? \text{ seconds} \\ v &= u - gt \\ 60 &= 120 - 9{\cdot}81\, t \\ \therefore\quad t &= 6{\cdot}12 \text{ seconds.} \end{aligned}$$

(ii) When velocity is 60 m/s moving downward,

$$\begin{aligned} \text{i.e.}\quad v &= -60 \text{ m/s} \\ u &= +120 \text{ m/s} \\ t &= ? \text{ seconds} \\ v &= u - gt \\ -60 &= 120 - 9{\cdot}81t \\ t &= 18{\cdot}35 \text{ seconds.} \end{aligned}$$

(*b*) Time to reach maximum height.
In this case the velocity v will be zero.

$$\begin{aligned} \therefore\quad 0 &= 120 - 9{\cdot}81t \\ t &= 12{\cdot}23 \text{ seconds.} \end{aligned}$$

(*c*) Maximum height reached.

$$\begin{aligned} v &= 0 \text{ m/s} \\ u &= +120 \text{ m/s} \\ s &= ? \text{ m} \\ v^2 - u^2 &= -2gs \\ 0 - 120^2 &= -2 \times 9{\cdot}81 \times s \\ s &= 733{\cdot}9 \text{ metres.} \end{aligned}$$

Alternatively:

$$\begin{aligned} s &= ut - \tfrac{1}{2}gt^2 \qquad t = 12{\cdot}23 \text{ seconds from } (b) \\ s &= (120 \times 12{\cdot}23) - (\tfrac{1}{2} \times 9{\cdot}81 \times 12{\cdot}23^2) \\ &= 733{\cdot}9 \text{ metres.} \end{aligned}$$

(*d*) Time of flight or time to reach ground again.

This time will be twice the time taken to reach the maximum height,

i.e. $2 \times 12{\cdot}23 = 24{\cdot}46$ seconds.

There is an alternative method to determine the time of flight. Neglecting the effect of air resistance, we can state that the velocity of the body on striking the ground will be equal, but opposite, to its initial velocity.

In this case initial velocity $u = +120$ m/s
final velocity $v = -120$ m/s
$t = ?$

Using expression
$$v = u - gt$$
$$-120 = +120 - 9{\cdot}81$$
$$= 24{\cdot}46 \text{ seconds.}$$

Velocity is 60 m/s after 6·12 and 18·35 seconds.
Time taken to reach maximum height is 12·23 seconds.
Maximum height reached is 733·9 metres.
Time taken to reach ground is 24·46 seconds.

Example 5.4

A body is projected vertically upwards with a velocity of 25 m/s from the top of a tower 30 m high. How long will it take the body to reach the ground at the foot of the tower?

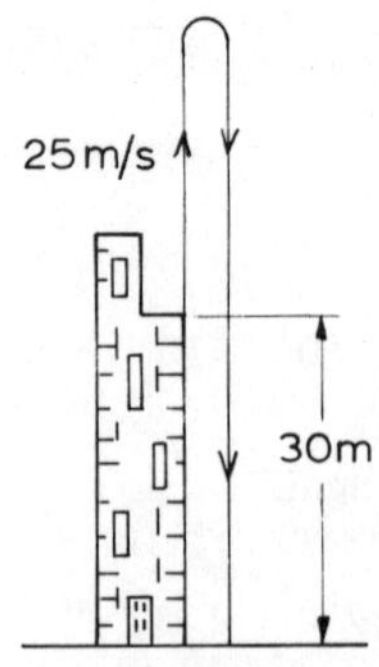

Figure 5.7

The main difficulty in a question of this type is that of the signs. There is also the question of the value of s in the formula $s = ut - \frac{1}{2}gt^2$. It is important to remember that s gives the distance measured from starting-point to finishing-point. It will not give the total distance travelled by the body. Its value will be -30 m, because the body is 30 m from the starting point when it comes to rest on the ground. The minus sign is introduced because downward distances and velocities are taken as negative.

$$u = +25 \text{ m/s}$$
$$s = -\ 30 \text{ m}$$
$$t = ?$$
$$s = ut - \tfrac{1}{2}gt^2$$

$$-30 = 25t - (\tfrac{1}{2} \times 9{\cdot}81 \times t^2)$$
$$4{\cdot}9t^2 - 25t - 30 = 0$$
$$t = 6{\cdot}1 \text{ seconds (the negative answer not being considered).}$$

The body reaches the foot of the tower after 6·1 seconds.

Example 5.5

A projectile is fired in a horizontal direction with a velocity of 500 m/s from the top of a cliff 100 m high. What will be the horizontal distance from the base of the cliff to the point where the projectile hits the sea?

The time of flight will be the same as that of a body falling vertically from rest through a distance of 100 metres.

$$s = ut - \tfrac{1}{2}gt^2 \qquad s = -100 \text{ m}$$
$$-100 = 0 - \tfrac{1}{2} \times 9{\cdot}81t^2 \qquad u = 0 \text{ m/s}$$
$$t = \sqrt{\frac{100}{4{\cdot}905}} = 4{\cdot}52 \text{ seconds.} \qquad t = ? \text{ seconds}$$

The projectile is moving forward with a horizontal velocity of 500 m/s for 4·52 seconds.

$$\text{Horizontal distance covered} = \text{velocity} \times \text{time}$$
$$= 500 \times 4{\cdot}52$$
$$= 2260 \text{ metres.}$$

The projectiles strikes the sea 2260 metres from the base of the cliff.

Example 5.6

A jet of water (Fig. 5.8) issues in a horizontal direction from the side of a tank. It is found that the path of the jet is such that in a horizontal distance of 500 mm the water falls 55·5 mm. Determine the velocity of the water as it left the tank.

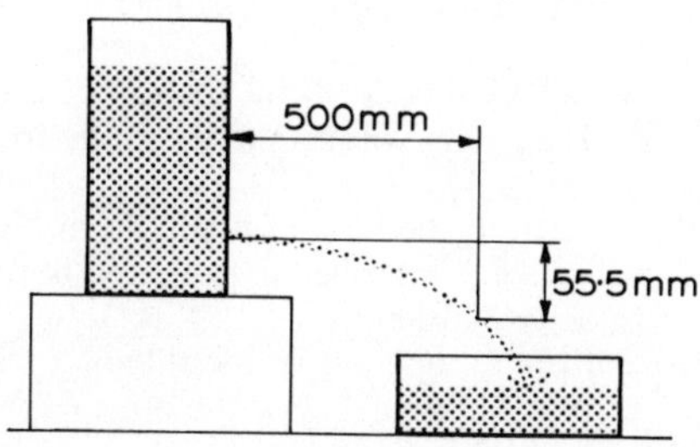

Figure 5.8 *Determination of Initial Velocity of Jet of Water*

Let v m/s = velocity of water on leaving tank.
Let t seconds = time to travel the 50 cm horizontally.

Considering the horizontal motion

$$vt = \tfrac{500}{1000} \text{ (converted into metres).}$$

Also considering the vertical motion, the water falls 0·0555 m from rest in seconds.

$$s = 0{\cdot}0555 \text{ m} \qquad s = ut - \tfrac{1}{2}gt^2$$
$$u = 0 \text{ m/s} \qquad 0{\cdot}0555 = 0 - \tfrac{1}{2} \times 9{\cdot}81 \times t^2$$
$$t = ? \text{ seconds} \qquad t = 0{\cdot}106 \text{ seconds.}$$

Now

$$vt = 0{\cdot}5$$
$$\therefore \quad v = \frac{0{\cdot}5}{t} = \frac{0{\cdot}5}{0{\cdot}106}$$
$$v = 4{\cdot}72 \text{ m/s.}$$

The jet of water has an initial horizontal velocity of 4·72 m/s on leaving the side of the tank.

EXERCISE 5

1 A stone, dropped from the top of a tower, strikes the ground after $2\frac{1}{2}$ seconds. Calculate the height of the tower and the velocity of the stone as it strikes the ground.

2 A projectile is fired vertically upwards with a velocity of 500 m/s. Neglecting air resistance, determine the height to which it will rise and the time after leaving the gun when it will be at 2000 m.

3 A rifle bullet is fired vertically upwards with a velocity of 800 m/s. Neglecting the effects due to air resistance, determine the height to which the bullet will rise and the time taken to reach the ground again.

4 A stone is allowed to fall from the edge of a cliff 100 m high. What will be its velocity after 1 second and after 3 seconds? How far will it have travelled at the end of 1 second and at the end of 3 seconds? How long will it take the stone to reach the base of the cliff and what will be its striking velocity?

5 A body is projected upwards with an initial velocity of 30 m/s. Find:
(*a*) the time when its velocity is 10 m/s,
(*b*) the time taken to reach the maximum height,
(*c*) its maximum height.

6 A stone is thrown vertically upward from the top of a tower 25 m high with a velocity of 20 m/s. How long will it take the stone to reach the base of the tower?

7 Determine the distance from the base of a cliff 45 m high to the point in the sea where a shell will strike, if the shell is projected horizontally from the top of the cliff with a velocity of 1000 m/s.

8 A jet of water issues from an opening in the side of a water tank with a horizontal velocity of 15 m/s. How far below the opening will the jet of water be at a horizontal distance of 2·5 m?

9 The path of a jet of water coming from an opening in the side of a water tank is such that the jet falls through a distance of 100 mm in a horizontal distance of 1 metre. What is the velocity (assumed horizontal) with which the jet leaves the tank?

Chapter Six
Angular Motion

Angular displacement

The engineer has no desire to be awkward when he decides to work in units of metres per second, whereas most people think about speed in terms of kilometres per hour. His excuse is that metres and seconds are considered to be the fundamental units of distance and time. You must not be surprised, therefore, to find that the engineer, in dealing with problems connected with rotation, i.e. with angular motion, decides to work in units of radians per second, whereas most people think about speed in terms of revolutions per minute. Again, his excuse is that the radian is considered to be the fundamental unit of angular displacement.

The radian

This is the angle between two radii drawn from the end of an arc whose length is equal to that of each of the radii. There are 2π radians in a complete circle, the size of a radian being $\frac{360°}{2\pi}$ or 57° 18′.

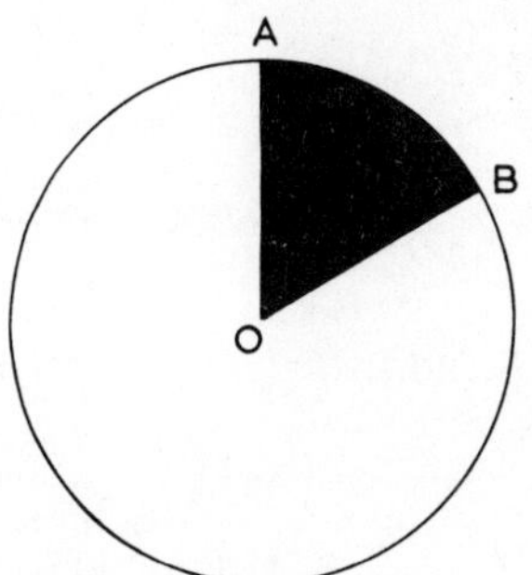

Figure 6.1 *The Radian*

The length of the arc AB measured around the circumference is equal to the radius of the circle OA. Angle AOB is then equal to 1 radian, which is approximately 57°. The radian is the basis of measurement of angular displacement.

Angular velocity

The line OA (Fig. 6.2) is rotating about the point O and has moved through an angle θ radians from the vertical position in t seconds. Its average angular velocity is $\frac{\theta}{t}$ radians per second. This angular velocity may be constant or it may vary. If it is constant, then the line OA passes through equal angles in equal intervals of time. We use the Greek letter omega (ω) to denote angular velocity and express its value in radians per second, abbreviated to rad/s.

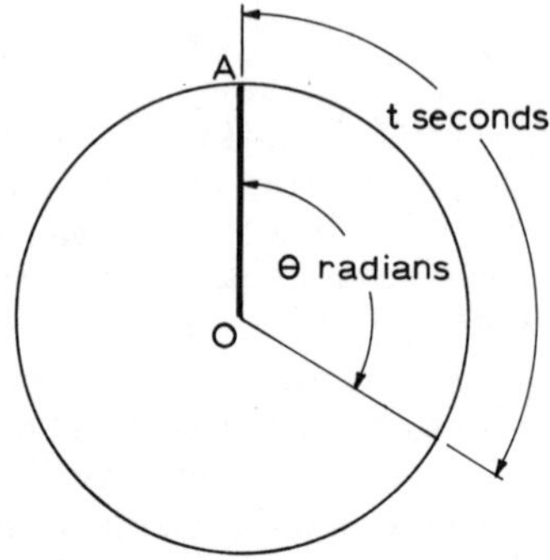

Figure 6.2 *Measurement of Angular Velocity*
The line OA moves through an angle of θ radians in t seconds. OA is moving with an angular velocity of θ/t radians per second.

We shall often want to convert revolutions per minute into radians per second.

$$N \text{ rev/min} = \frac{N}{60} \text{ rev/s}$$

$$= 2\pi \times \frac{N}{60} \text{ rad/s (since 1 revolution} = 2\pi \text{ rad)}$$

$$= \frac{\pi}{30} N \text{ rad/s.}$$

Also
$$\omega \text{ rad/s} = \frac{30\omega}{\pi} \text{ rev/min.}$$

Angular acceleration

If the angular velocity is not constant, then the line OA is either accelerating or retarding. Angular acceleration is an indication of the rate at which the angular velocity is changing. If, for example, the angular velocity of the line OA about the point O is:

4 rad/s at the end of 1st second,
8 rad/s at the end of 2nd second,
12 rad/s at the end of 3rd second, etc.

then the angular velocity is increasing at the rate of 4 rad/s during every second. We say that the angular acceleration is 4 rad/s per second or 4 rad/s^2. We use the Greek letter alpha (α) to denote angular acceleration.

DEFINITION. **Angular acceleration is the rate of change of angular velocity.**

Conversion from angular to linear velocity

The wheel in Fig. 6.3 is rotating at ω rad/s. Any line such as OA, length r m, will have turned through ωt radians in t seconds. The distance s travelled by the point A on the rim of the wheel is given by the expression:

$$\text{length of arc} = \text{angle in radians} \times \text{radius}$$
$$s = \omega t \times r$$
$$s = \omega r t \text{ m}$$

i.e. A travels a distance $\omega r t$ m in t seconds. Its linear velocity is

$$\frac{\omega r t}{t} = \omega r \text{ m/s.}$$

$$\text{Angular velocity} = \omega \text{ rad/s}$$
$$\text{Linear velocity } v = \omega r \text{ m/s}$$
$$\text{Angular acceleration} = \alpha \text{ rad/s}^2$$
$$\text{Linear acceleration } f = \alpha r \text{ m/s}^2$$

Hence to find the velocity of any point on a rotating arm or wheel, multiply the angular velocity of the wheel by the distance of the point from

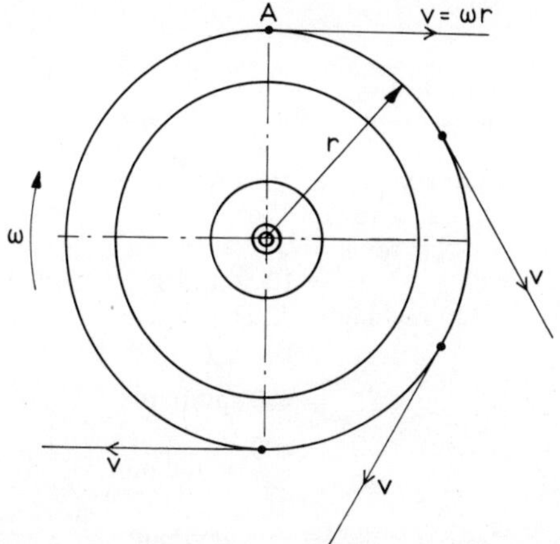

Figure 6.3 *Conversion from Angular to Linear Velocity*

The wheel is rotating with angular velocity ω rad/s. A point on the rim has a linear velocity $v = \omega r$ m/s, where r is the radius of the wheel in metres. Although the magnitude of the velocity of A is constant, its direction is always changing. The direction is always tangential to the rim. Consequently, since the direction of the velocity is changing, the point on the rim is accelerating.

the axis of rotation. *The direction of the velocity of the point is always perpendicular to the line joining the point to the axis of rotation.*

NOTE. We have already seen that the velocity of a point moving in a circle cannot be constant, because of the changing direction. The magnitude is constant, but not the direction.

Conversely, if we know the linear velocity of a point on a wheel, we can find the angular velocity of the wheel by the expression:

$$\text{Angular velocity} = \frac{\text{linear velocity (m/s)}}{\text{radius (m)}}$$

$$\omega = \frac{v}{r} \text{ rad/s.}$$

Similar expressions apply to angular and linear accelerations.

$$\text{Linear acceleration} = \text{angular acceleration} \times \text{radius}$$
$$f = \alpha r \text{ m/s}^2$$

$$\text{Angular acceleration} = \frac{\text{linear acceleration (m/s}^2\text{)}}{\text{radius (m)}}$$

$$\alpha = \frac{f}{r} \text{ rad/s}^2.$$

Example 6.1

Convert: (*a*) 120 rev/min into rad/s
(*b*) 8 rad/s into rev/min.

(*a*)

$$\begin{aligned} 120 \text{ rev/min} &= \tfrac{120}{60} \text{ rev/s} \\ 1 \text{ rev} &= 2\pi \text{ rad} \\ \tfrac{120}{60} \text{ rev/s} &= \tfrac{120}{60} \times 2\pi \text{ rad/s} \\ &= 12{\cdot}56 \text{ rad/s} \\ \mathbf{120\ rev/min} &= \mathbf{12{\cdot}56\ rad/s.} \end{aligned}$$

(*b*)

$$\begin{aligned} 8 \text{ rad/s} &= 8 \times 60 \\ &= 480 \text{ rad/min} \\ &= \frac{480}{2\pi} \times \text{rev/min} \\ &= 76{\cdot}5 \text{ rev/min} \\ \mathbf{8\ rad/s} &= \mathbf{76{\cdot}5\ rev/min.} \end{aligned}$$

Example 6.2

The diameter of the wheels of a vehicle travelling at 30 km/h is 1·2 m. What is the angular velocity of the wheels?

$$30 \text{ km/h} = \frac{30 \times 1000}{60 \times 60}$$
$$= 8{\cdot}33 \text{ m/s}.$$

This is the linear velocity of a point on the rim of the wheel.

$$\text{Angular velocity} = \frac{\text{linear velocity (m/s)}}{\text{radius (m)}}$$
$$= \frac{8{\cdot}33}{0{\cdot}6}$$
$$= 13{\cdot}9 \text{ rad/s}$$
$$= \frac{13{\cdot}9 \times 60}{2\pi} \text{ rev/min}$$
$$= 132 \text{ rev/min}.$$

Angular velocity of wheels is 13·9 rad/s or 132 rev/min.

Angular displacement, velocity, acceleration and time

ω_1 = initial velocity in rad/s
ω_2 = final velocity in rad/s after t seconds
α = angular acceleration in rad/s^2
θ = angle turned through in radians
t = time in seconds.

The initial angular velocity ω_1 will be increased by α rad/s at the end of each second. At the end of t seconds its velocity will be $\omega_1 + \alpha t$.

This gives:

$$\omega_2 = \omega_1 + \alpha t \qquad (1)$$

or

$$\omega_2 - \omega_1 = \alpha t \qquad (2)$$

Since the velocity is increasing uniformly, the average velocity is $\dfrac{\omega_1 + \omega_2}{2}$, and since:

$$\text{displacement} = \text{average velocity} \times \text{time}$$
$$\theta = \frac{\omega_1 + \omega_2}{2}t \qquad (3)$$

Rearranging, we have:

$$(\omega_1 + \omega_2) = \frac{2\theta}{t}$$

$$(\omega_2 + \omega_1) = \frac{2\theta}{t} \qquad (4)$$

Multiplying (2) and (4) together:

$$(\omega_2 + \omega_1)(\omega_2 - \omega_1) = \frac{2\theta}{t} \times \alpha t.$$

The left-hand side is the difference of two squares, and on the right-hand side we can cancel out the t's. Hence we get:

$$\omega_2{}^2 - \omega_1{}^2 = 2\theta\alpha \qquad (5)$$

Again substituting the value of ω_2 as given in equation (1) into equation (3):

$$\theta = \frac{[\omega_1 + (\omega_1 + \alpha t)]t}{2}$$

$$\theta = \frac{[2\omega_1 + \alpha t]t}{2}$$

$$\theta = \omega_1 t + \tfrac{1}{2}\alpha t^2 \qquad (6)$$

Notice the similarity with those obtained for linear motion.

Example 6.3

An electric motor is rotating at 3000 rev/min when it is brought to rest with uniform retardation in 15 seconds.

(*a*) Calculate the value of this retardation.
(*b*) Calculate the number of revolutions made by the motor in coming to rest.

$$\text{Initial velocity} = 3000 \text{ rev/min}$$
$$= \frac{3000}{60} \times 2\pi \text{ rad/s}$$
$$= 314{\cdot}6 \text{ rad/s}$$
$$\text{Final velocity} = 0 \text{ rad/s (at rest).}$$

(*a*) Change in velocity = final − initial velocity

$$= 0 - 314{\cdot}6$$
$$= -314{\cdot}6 \text{ rad/s}$$

$$\text{Acceleration} = \frac{\text{change in velocity}}{\text{time}}$$
$$= \frac{-314{\cdot}6}{15}$$
$$= -21 \text{ rad/s}^2.$$

The negative sign implies a retardation.

(*b*) Average velocity during retardation

$$= \frac{\text{initial} + \text{final velocity}}{2}$$
$$= \frac{0 + 314{\cdot}6}{2}$$
$$= 157{\cdot}3 \text{ rad/s}.$$

Angle turned through in 15 seconds at average velocity of 157·3 rad/s

$$= \text{velocity} \times \text{time}$$
$$= 157{\cdot}3 \times 15$$
$$= 2360 \text{ radians.}$$

$$\text{Revolutions} = \frac{2360}{2\pi}$$
$$= 375.$$

The motor has a retardation of 21 rad/s² and makes 375 revolutions in coming to rest.

NOTE. An alternative solution to part (*b*) would be as follows:

$$3000 \text{ rev/min} = 50 \text{ rev/s.}$$

$$\text{Average velocity} = \frac{0 + 50}{2}$$
$$= 25 \text{ rev/s.}$$

$$\text{Revolutions in 15 seconds} = \text{velocity} \times \text{time}$$
$$= 25 \times 15$$
$$= 375 \text{ revolutions.}$$

EXERCISE 6

1 Convert: (*a*) 300 rev/min into rad/s.
(*b*) 5 rev/min into rad/s.
(*c*) 50 rad/s into rev/min.
(*d*) 180 rad/s into rev/min.

2 Convert into rad/s² the following angular accelerations:
(*a*) 4 rev/min².
(*b*) 100 rev/min per second.

3 What is the angular velocity of the 2m diameter wheels of a locomotive engine which is travelling at 100 km/h?

4 A wheel is rotating at 100 rev/min. What is the velocity in m/s of a point on the rim which is 1 m from the centre of the wheel?

5 Two pulleys A and B are connected by a belt. Pulley A is 1 m diameter and runs at 200 rev/min. Pulley B is 0·75 m diameter. What is the linear velocity of the belt in m/s? Hence determine the angular velocity of B expressed in rad/s and rev/min.

6 There is a rule used by millwrights in connection with the design of cast-iron wheels to prevent the 'bursting' of the casting when rotating. The rule states that the speed of a point on the rim of the wheel must not exceed 100 kilometres per hour.

What would be the limiting speed of (*a*) a 1·2 m diameter flywheel, (*b*) a rope pulley 2·2 m diameter?

7 A flywheel is rotating at a uniform speed of 150 rev/min. If it is brought to rest in 10 min, determine the angular retardation expressed in rad/s². How many revolutions does the wheel make in coming to rest?

8 A train is travelling at 15 km/h. The train wheels are 1 m in diameter. What is the angular velocity of the wheels? If the train is brought to rest in 30 m, what is the retardation of the train in m/s^2 units? What is the angular retardation of the wheels?

9 The armature of an electric motor is rotating at 5000 rev/min. The motor comes to rest in 10 seconds after the current is switched off. Calculate:
(*a*) the average angular retardation of the armature.
(*b*) the number of revolutions made by the armature after the current is switched off. [U.E.I.]

10 A car with wheels 750 mm in diameter moving at 25 km/h is brought to rest under a uniform retardation in a distance of 60 m. Find:
(*a*) the retardation of the car.
(*b*) the initial angular velocity and the angular retardation of the wheels.
State clearly the units in which each answer is expressed. [U.E.I.]

11 A vehicle with wheels 1 m in diameter has its speed reduced uniformly from 50 km/h to 25 km/h while it travels a distance of 12 m. Determine:
(*a*) the time taken in seconds,
(*b*) the linear retardation of the vehicle in m/s^2,
(*c*) the angular retardation of the wheels in rad/s^2. [U.L.C.I.]

12 A flywheel is uniformly accelerated from rest and reaches 30 rev/min in 2 min. Find:
(*a*) the acceleration in rad/s^2,
(*b*) the number of revolutions made by the flywheel when the speed has reached 120 rev/min. [N.C.T.E.C.]

13 Establish the relation between the angular speed ω of a rotating disc and the linear speed, v, of a point on it at distance r from the centre of rotation.
A disc 1 m in diameter rotates at 150 rev/min. What is the angular speed in rad/s, and what is the speed of the rim in m/s? [N.C.T.E.C.]

14 A winch drum, 1 m in diameter is paying out rope at a constant speed of 12 m/s when the brake is applied. If the winch comes to rest in 10 seconds and the deceleration may be assumed uniform, determine: (*a*) the initial angular velocity of the drum, (*b*) the angular deceleration, and (*c*) the number of revolutions made by the drum in the 10 seconds before coming to rest. [N.C.T.E.C.]

15 A winding drum of 8 m diameter starts from rest with an angular acceleration of 0·5 rad/s^2. After 30 seconds of acceleration the speed remains constant for 1 min, the drum then being brought to rest in a further 20 seconds. Determine:
(*a*) the maximum rim speed of the drum in m/s,
(*b*) the total number of revolutions made, and
(*c*) the deceleration during the final stage. [N.C.T.E.C.]

Chapter Seven
Force and Acceleration

When a force is applied to a body at rest, but which is free to move, *the force causes the body to accelerate.* In the same way, a force applied to a moving body will cause a change in the velocity, either an increase or a decrease, depending upon the nature and direction of the applied force. When the brakes are applied to a car, the forces exerted by the brakes cause the car to slow down. The thrusts provided by the various jets on a spacecraft are used to increase or decrease the speed of the spacecraft as well as to provide changes in direction.

The question which we have to answer is: What is the relationship between the applied force and the resulting acceleration? Upon what other factor, if anything, is the resultant acceleration dependent? We know from experience that it is much easier to start an empty truck moving than it is to start a full truck moving. Once the trucks are moving, the forces required to maintain uniform motion will be practically the same. We shall require a much larger force to stop the full truck than to stop the empty truck in the same time. The force to produce either the acceleration or the retardation is dependent upon the 'quantity of matter' which is to be accelerated.

DEFINITION. **The mass of a body is the quantity of matter which it contains.**

Gravitational forces between bodies—weight

We are all familiar with the attraction which exists between a steel magnet and a piece of steel. This attraction is due to magnetic forces.

Now there is a force of attraction (due to causes other than magnetic) existing between all bodies. This force is usually called a gravitational force. It is present between all bodies, whether they be magnetic or not. The magnitude of the force depends upon the mass of the two bodies and the distance between them. Sir Isaac Newton discovered the law relating the force to the masses and the distance between them and expressed it as follows:

$$\text{Force} = \text{constant} \times \frac{\text{product of the two masses}}{(\text{distance between them})^2}.$$

That is, the force will be greater between larger masses than between smaller masses, and will also be greater between bodies near together than between those farther apart. Between two books on the desk, there is a force of attraction. Between a hammer and a file on the bench, there is a force of attraction. These forces are, of course, very small indeed,

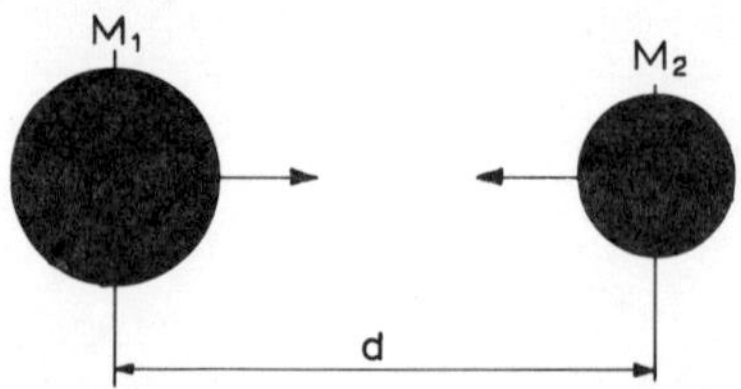

Figure 7.1 *Attraction between Two Bodies*

A force of attraction exists between all bodies. This force, which is normally very small, depends upon the mass of each body and the distance between them. When one of the bodies is the earth, then the force can be very considerable. If the other body is free to move, it will be drawn towards the earth. In other words, it will fall. The gravitational force between the earth and the body is known as the *weight* of the body.

far too small to cause the books to come together or the hammer and the file to come together. An idea of the magnitude of the force of attraction can be obtained when we think of two 20 000 tonne ships placed side by side 100 metres apart. The pull between the two is approximately 35 N. When we think of much greater masses such as the sun and the earth and the moon, we find that there are some extremely large attractive forces between them. For example, the attraction between the earth and the sun causes the earth to be deflected from the straight path which it would normally take, and causes it to move along an elliptical path around the sun. The force of attraction between the moon and the oceans of the earth is so great as to cause the actual movement of the ocean. It is drawn towards the moon. This causes the tides to ebb and flow.

The force of attraction between the earth and the hammer which you may be holding is sufficiently great to make you apply an upward force to prevent the hammer from dropping to the floor. If you stop applying the upward force, the hammer will fall towards the centre of the earth, pulled by the force of gravity.

Now this gravitational force which is exerted upon the body is sometimes called the **weight** *of the body*, and it depends, as we have seen, upon:

the mass of the hammer;
the mass of the earth, and
the distance between the centre of the hammer and the centre of the earth.

Since the mass of the hammer and of the earth are both constant, we can vary the force of attraction or the weight of the hammer if we increase the distance between the bodies.

Increase the distance – decrease the weight
Decrease the distance – increase the weight.

Therefore, the hammer will weigh more at the bottom of a mine than it will in an aeroplane at 10 000 m. The hammer will not weigh as much at the Equator as it does at the Poles, because, as the earth is shaped like an orange, places on the Equator are farther from the centre than places near to the Poles. Of course, the variation in the weight will only be small, but it does, in fact, exist.

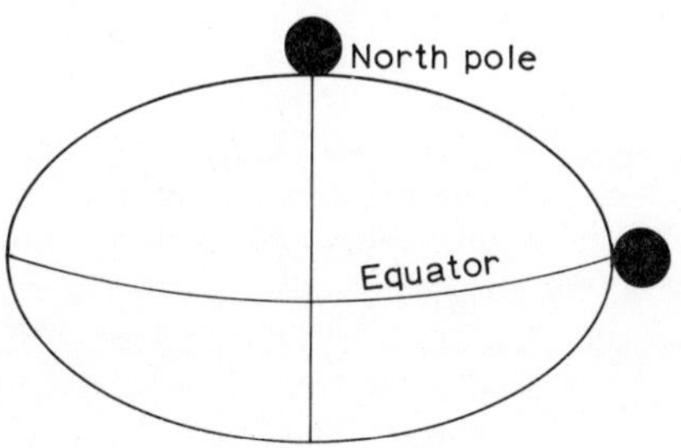

Figure 7.2 *The Changing Weight of a Body*

The weight of a body is the force which the earth exerts on the body. This force depends upon the mass of the earth, the mass of the body and the distance between the two. Since the earth is not perfectly spherical (but not as oval as the exaggerated diagram above), the body will be nearer to the centre of the earth at the Poles than at the Equator. Consequently, the force on the body will be greater at the Poles than at the Equator. All of which means that the body will weigh more at the Poles than at the Equator.

If the hammer could be weighed on the moon, it would weigh considerably less than it does on the earth. The mass of the moon is much smaller than that of the earth; consequently, the force of gravity which it exerts on the hammer must be smaller than the corresponding force exerted by the earth on the hammer.

It is very important that we realise the difference between mass and weight.

The mass of a body is constant.
The weight of the body varies from place to place.

Measurement of force and mass

There are two usual methods of 'weighing' a body. One is by the beam balance, in which certain 'weights' of known magnitude are put in one scale pan to balance the body placed in the other scale pan. The other method is to use some type of spring balance in which the extension or compression of the spring, to which the body is attached, indicates the 'weight' of the body.

The beam balance only compares masses. It simply indicates that the quantity of matter in the body is equal to the quantity of matter in the 'weights'. Once you have reached a balance, it will be maintained wherever you may be. If they balance in London they will balance on the moon, because the variation in the force of gravity affects both the 'weights' and the body which you are 'weighing.'

On the other hand, the spring balance measures forces. It measures the pull exerted by the earth on the body; it measures the weight of the body. Your spring balance will show a smaller reading on the moon that it does in London. *The spring balance is the method of obtaining the true weight of the body.*

As an experiment, try to weigh something with a spring balance while you are travelling in a lift. When you start to go up, the spring balance reading will increase. When you are moving with constant velocity, it will return to its normal figure. When you slow down to stop, the reading will decrease. When you have stopped, the reading will again be normal. But if you carried out this experiment using a beam balance, you will find that the balance is unaffected whether you are going up or coming down. You see, both sides of the balance are affected in the same way.

The spring balance is used for obtaining the true weight. (Of course, you ought not to use it in a moving lift.)

The beam balance simply compares masses.

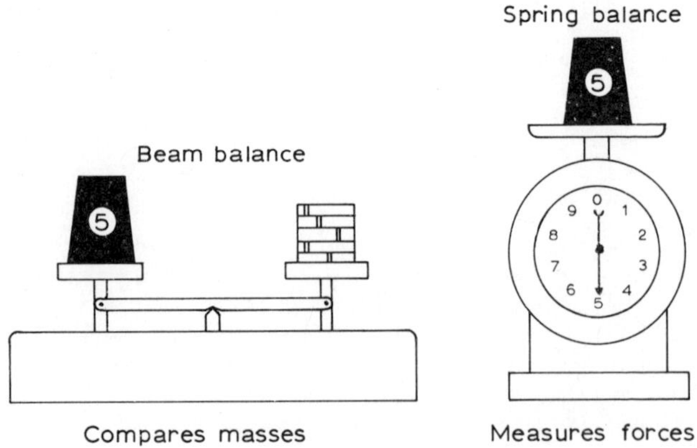

Figure 7.3

There is a standard mass of platinum in the Treasury Office in London and a standard mass of 1 kilogram in the laboratories of the International Bureau of Weights and Measures at Sèvres, outside Paris. Certain sub-standard masses can be compared with these standards. Ultimately, the weights which the shopkeeper uses are compared with the sub-standards so that in effect the substance which you buy in the shop has its mass compared with the official standard.

Momentum

DEFINITION. **The momentum of a body is the product of its mass and its velocity.**

Momentum is a vector quantity, since, involving velocity, it involves direction. It can be resolved into components, or compounded with other momenta (which is the plural of momentum) in the same way that we deal with forces and velocities.

The momentum of a body is changed if we change either the mass or the velocity of the body.

Change of momentum due to change of mass:

The *mass* of a rocket in flight is decreasing as the fuel burns away.

The *mass* of a water cart in the street is decreasing as the water leaks away.

The *momentum* of the rocket and the water cart is decreasing.

Change of momentum due to change of velocity:

The momentum of a truck on the Big Dipper increases as it speeds down the straight track and then decreases as it shoots up the other side. (The magnitude of the velocity is changing in this case.)

The momentum of a jet of water changes as it strikes against a wall. (The direction of the velocity is changing in this case; after striking the wall, the direction of the water is changed.)

Whenever a change in momentum takes place, there is always a force either causing the change or being caused by the change.

In his second law of motion, Sir Isaac Newton stated that:

The rate of change of momentum is proportional to the applied force and takes place in the direction of the applied force.

Force $\propto$ rate of change of momentum
($\propto$ signifies for 'is proportional to')
$\propto$ rate of change of mv
$\propto$ $m \times$ rate of change of v if m is constant
$\propto$ mass $\times$ acceleration, since rate of change of velocity is equal to acceleration.

Writing this in the form of an equation we have:

$$\text{Force} = k \times \text{mass} \times \text{acceleration},$$

where k is a constant.

We can make the constant equal to 1 *if we choose a force which will give a unit mass a unit acceleration.*

The unit force chosen to satisfy this condition is the *newton* (N).

The newton is the force which, acting upon a mass of 1 kilogram, will cause it to move with an acceleration of 1 metre per second per second.

An idea of the magnitude of one newton can be obtained by holding a mass of one quarter of a pound. The gravitational force exerted on the weight is approximately one newton. The accurate relationship would be that a force of one newton is equal to the gravitational force acting on a mass of 0·225 lb.

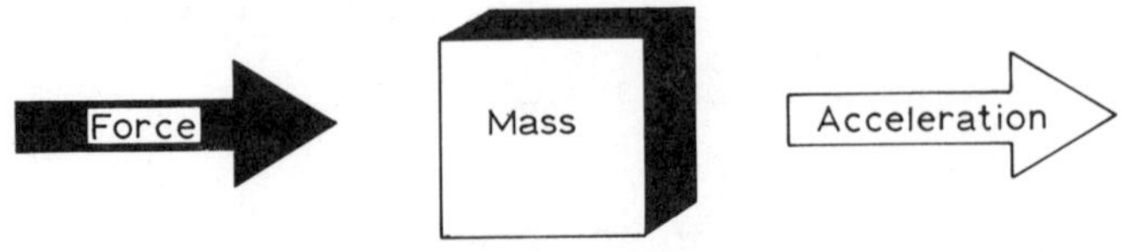

Figure 7.4

The acceleration produced is calculated from the fundamental relationship: Force = mass × acceleration.

We have already seen that a mass falling freely has an acceleration of g m/s^2. The acceleration is caused by the gravitational force acting on the mass. For the purpose of calculation, an average value of 9·81 m/s^2 is used.

$$\text{Gravitational force (newtons)} = \text{mass (kilograms)} \times g$$
$$= 9{\cdot}81 \text{ N.}$$

So if you were to place a mass of 1 kilogram on a spring balance calibrated in newtons, you would find that the spring balance indicated a force of 9·81 newtons.

Two points come to mind here.

1 One is tempted to comment that the weight of a mass of 1 kilogram is 9·81 newtons. The use of the expression 'weight' has been commonplace, albeit confusing, for as long as records have been kept. With the current move into an age of SI units and interplanetary travel, the term 'weight' should become extinct. Whether or not this will be so remains to be seen.

2 Some people have already decided to use a figure of 10 rather than 9·81 in their engineering calculations. It is argued that in some aspects of engineering, a design based on 10 will be acceptable since it is within 2% of the more accurate figure. Calculations are certainly simplified, and the use of 10 is in keeping with other calculations which metrication has brought. Its use must be left to the discretion of the engineer. Perhaps it can be used as a convenient checking factor.

Example 7.1

A car has a mass of 1000 kg. Determine the acceleration of the car when a force of 150 N moves it over a horizontal road.

$$\text{Mass} = 1000 \text{ kg}$$
$$\text{Force} = 150 \text{ N.}$$
$$\text{Force (newtons)} = \text{mass (kg)} \times \text{acceleration}$$

$$150 = 1000 \times f$$
$$f = \frac{150}{1000}$$
$$f = 0{\cdot}15 \text{ m/s}^2.$$

Acceleration of the car is 0·15 m/s².

Example 7.2

A man and his cycle have mass of 100 kg. When travelling at 25 km/h on a level road, the cyclist suddenly ceases to pedal, and in 25 seconds finds that his speed has changed to 15 km/h. What is the total force resisting motion?

$$\text{Initial velocity} = \frac{25 \times 1000}{60 \times 60} = 6{\cdot}95 \text{ m/s}$$

$$\text{Final velocity} = \frac{15 \times 1000}{60 \times 60} = 4{\cdot}16 \text{ m/s}$$

$$\begin{aligned} \text{Change in velocity} &= \text{final} - \text{velocity} \\ &= 4{\cdot}16 - 6{\cdot}95 \\ &= -2{\cdot}78 \text{ in 25 seconds.} \end{aligned}$$

$$\begin{aligned} \text{Acceleration} &= \frac{\text{change in velocity}}{\text{time}} \\ &= \frac{-2{\cdot}78}{25} \text{ m/s}^2 \\ &= -0{\cdot}111 \text{ m/s}^2 \text{ (negative sign indicates retardation).} \end{aligned}$$

The mass of the man and the cycle is 100 kg.

$$\begin{aligned} \text{Force (newtons)} &= \text{mass (kg)} \times \text{acceleration} \\ &= 100 \times 0{\cdot}111 \\ &= 11{\cdot}1 \text{ newtons} \end{aligned}$$

The force resisting motion is 11·1 newtons.

NOTE. The negative sign in connection with the acceleration is not brought into the second part of the solution. The point is that a force of 11·1 newtons would be required to accelerate the man and cycle in the same way that a force of 11·1 newtons would be required to retard them. The essential difference is that in the one case the force acts in the direction of motion, in the other case it opposes motion.

Example 7.3

A mass of 100 kg is raised vertically by a rope. The load reaches a speed of 25 m/s from rest with uniform acceleration after rising 20 m. Determine the tension in the rope.

Acceleration of the load

$$\begin{aligned} v^2 - u^2 &= 2fs \\ 25^2 - 0^2 &= 2 \times 20f \\ f &= 15{\cdot}6 \text{ m/s}^2 \end{aligned}$$

Tension in the rope $= T$ newtons.

The gravitational force acting on the 100 kg mass is equal to 100 g newtons, i.e. 981 newtons.

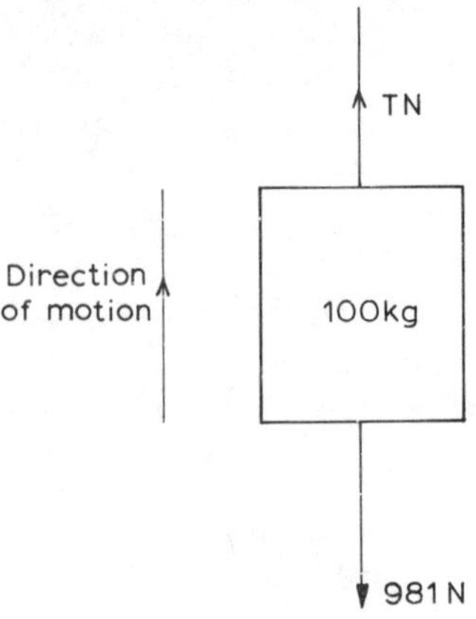

Figure 7.5

Forces acting on load (1) T newtons tension in rope up.
(2) 981 newtons gravitation pull down.

Resultant upward force = $(T - 981)$ N
Mass moving = 100 kg
Force (newtons) = mass (kg) × acceleration
$(T - 981) = 100 \times 15{\cdot}6$
$T = 2541$ newtons

The tension in the rope is 2541 newtons.

Example 7.4

In a straight length of 150 m, a road falls uniformly through a height of 2 m. A cyclist with his brakes off just freewheels at a uniform speed down the hill until he comes on to the flat. With what horizontal force must he now propel his machine to maintain the same speed? The cyclist and machine together have a mass of 100 kg. [N.C.T.E.C.]

Weight of cyclist and machine = 100 g
= 981 N.

The force moving the cyclist down the incline is $981 \sin \theta$, where θ is the angle of the incline.

$$\text{Sin } \theta = \frac{2}{150}$$

$$\therefore \quad \text{Force moving down plane} = 981 \times \frac{2}{150}$$

$$= 13{\cdot}1 \text{ N.}$$

This force will be equal and opposite to the frictional resisting forces, which can be taken as constant.

Hence to move at a uniform speed along the level the cyclist must propel the cycle with a force of 13·1 N.

Example 7.5

A truck of mass 500 kilograms travelling with a velocity of 30 kilometres per hour is brought to rest in 30 seconds. What is the magnitude of the uniform retarding force?

$$30 \text{ km/h} = \frac{30 \times 1000}{60 \times 60}$$
$$= 8{\cdot}33 \text{ m/s.}$$

This velocity is reduced to zero in 30 seconds.

$$\therefore \quad \text{Retardation} = \frac{8{\cdot}33}{30}$$
$$= 0{\cdot}277 \text{ m/s}^2.$$
$$\text{Force} = \text{mass} \times \text{acceleration (retardation in this case)}$$
$$= 500 \times 0{\cdot}277$$
$$= 138{\cdot}5 \text{ newtons.}$$

Average force acting through retardation is 138·5 newtons.

Motion on an incline

If a car is moving on a slope rather than on the level, there is an additional force which we shall have to consider. If the mass of the car is M,

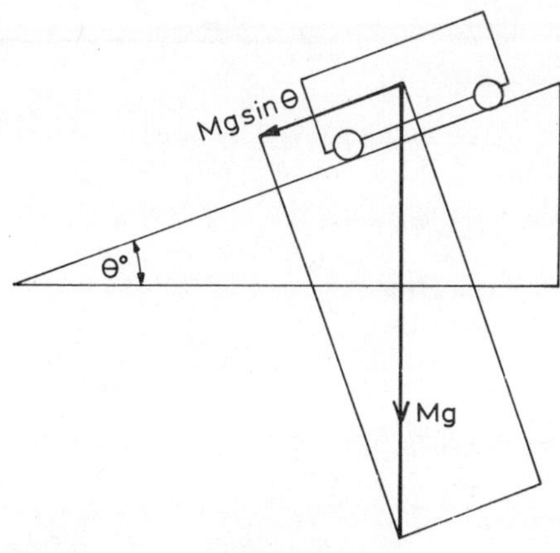

Figure 7.6

its weight will be Mg. If the angle of the slope is θ, then there is a force $Mg \sin \theta$ pulling the car down the slope. This is the component of the weight of the car down the plane.

If the car is travelling up the slope, then it must exert a force to:

1. overcome $Mg \sin \theta$ down the plane,
2. overcome general resistance, and
3. accelerate the car.

The force which the car can exert is called the *tractive effort* of the car. If the tractive effort of a car does not exceed 1. above, then the car will not be able to climb up the slope. In fact, the tractive effort must exceed the sum of 1. and 2. above. The amount by which it exceeds the sum of 1. and 2. will indicate how much force there is to accelerate the car up the slope.

If the car is moving down the slope, then the component of the weight of the car down the slope will be greater than the resistances and the car will gradually accelerate, the value of the acceleration depending on the amount by which the component exceeds the resistance. Two assumptions have been made in the last sentence. First, that the car is running downhill freely and not being driven by the engine. Second, that the value of the resistance is constant. This is not perfectly true and at a later stage you will deal with the way in which the resistances vary with the change in velocity.

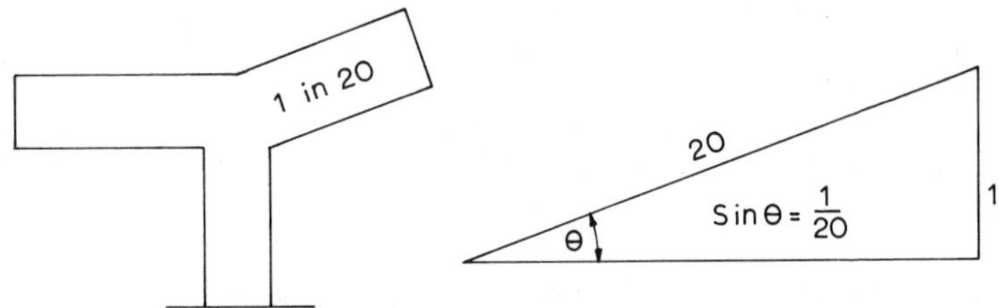

Figure 7.7 *Slope of Incline*

Incidentally, when we say that the slope of an incline is 1 in 20, we mean that by moving 20 m along the slope we move 1 m vertically. Hence the slope, expressed as a fraction, is equal to the sine of the angle of the slope.

Example 7.6

A car, mass 1000 kg, starting from rest climbs a gradient of 1 in 8 against a frictional resistance of 150 N. The tractive effort of the car is 2000 N. Determine its acceleration up the gradient. What would be its acceleration on the level?

$$\text{Weight of the car} = Mg = 1000 \times 9{\cdot}81 \text{ N.}$$

$$\text{Component of the weight of car down the plane} = 1000 \times 9{\cdot}81 \times \tfrac{1}{8}$$
$$= 1226 \text{ N.}$$

Car has to overcome gravitation effect plus frictional resistance, i.e. $1226 + 150 = 1376$ N.

$$\text{Hence force available for acceleration} = 2000 - 1376$$
$$= 624 \text{ N.}$$
$$\text{Mass of car} = 1000 \text{ kg}$$
$$\text{Acceleration} = \frac{\text{force (newtons)}}{\text{mass (kilograms)}}$$
$$= \frac{624}{1000}$$
$$= 0{\cdot}624 \text{ m/s}^2.$$

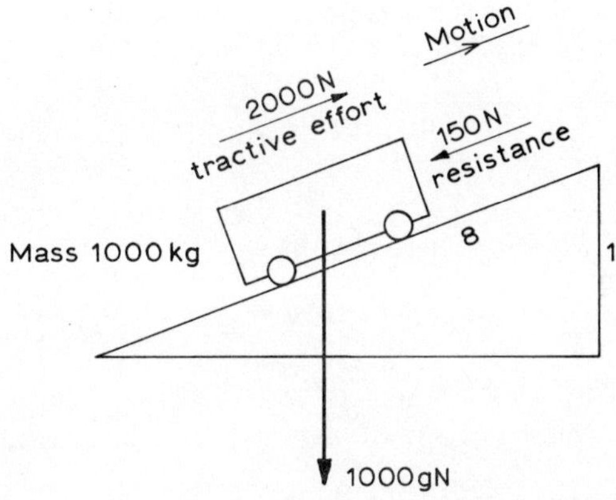

Figure 7.8

On the level, the acceleration force available is the tractive force less the frictional resistance.

$$\text{i.e.}\quad 2000 - 150 = 1850 \text{ N}$$

$$\therefore \text{Acceleration on level} = \frac{\text{force (newtons)}}{\text{mass (kilograms)}}$$

$$= \frac{1850}{1000}$$

$$= 1{\cdot}85 \text{ m/s}^2.$$

Acceleration up gradient is 0·624 m/s², whilst on the level it is 1·85 m/s².

Example 7.7

A car descends an incline of 1 in 20 with the engine switched off. The mass of the car is 4000 kg and the resistances to motion are equal to 200 N. Find the force causing the car to accelerate and the magnitude of the acceleration.

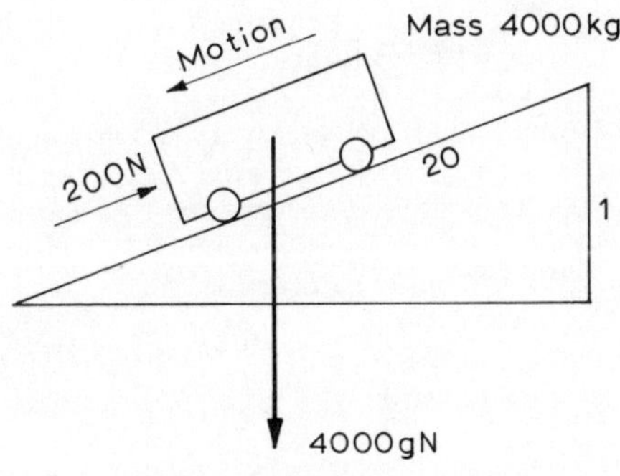

Figure 7.9

$$\text{Weight of car} = Mg = 4000 \times 9{\cdot}81 \text{ N.}$$

$$\begin{aligned}\text{Component of weight of car down incline} &= 39\,240 \text{ N} \times \text{slope}\\ &= 39\,240 \times \frac{1}{20}\\ &= 1962 \text{ N.}\end{aligned}$$

NOTE. If the resistances are greater than 1962 N, the car will not run freely down the incline—it will need the engine to drive it down.

$$\begin{aligned}\text{Force available to accelerate} &= 1962 - \text{resistances}\\ &= 1960 - 200\\ &= 1760 \text{ N.}\end{aligned}$$

This force acting on the mass of the car will cause it to accelerate.

$$\begin{aligned}\text{Mass of car} &= 4000 \text{ kg}\\ \text{Acceleration (m/s}^2\text{)} &= \frac{\text{force (newtons)}}{\text{mass (kilograms)}}\\ &= \frac{1760}{4000}\\ &= 0{\cdot}44 \text{ m/s}^2.\end{aligned}$$

The force causing acceleration is 1760 N, producing an acceleration of 0·44 m/s².

EXERCISE 7

Remember that the weight of a mass of M *kg is always* Mg *newtons.*

1 What force is required to cause a mass of 100 kg to move with an acceleration of 4 m/s²?

2 A block whose mass is 2 kg is moved by a force of 0·5 N. What is the resulting acceleration?

3 An engine exerts a pull of 20 kN on a train whose mass is 250 tonnes. How long will it take the train to reach a speed of 50 km/h from rest?

4 If the resistances to motion can be taken as 50 N per tonne, what will be the acceleration of a 300 tonnes train which is pulled by an engine exerting a draw-bar pull of 25 kN?

5 A force of 10 N pulls a 15 kg block over a horizontal plane and moves through a distance of 5 m in 6 seconds from rest. Determine the acceleration of the block, the force required to produce this acceleration, and consequently the frictional force opposing motion.

6 A car moving at 60 km/h along a horizontal road is brought to rest in 60 m. If the mass of the car is 1 tonne, what is the average retarding force?

7 Determine the tension in the rope of a lift whose mass is 500 kg when the lift is (*a*) accelerating upwards at 1 m/s², (*b*) accelerating downwards at 1 m/s².

8 What constant resistance will reduce the speed of a mass of 1000 kg from 50 km/h to 10 km/h in $1\frac{1}{2}$ min?

9 A car moves at 60 km/h on a horizontal road and is brought to rest in 60 m. The mass of the car is 1 tonne.
(*a*) What is the average retarding force?
(*b*) How long does the car take to come to rest? [U.E.I.]

10 A man and his cycle have a mass of 80 kg; when travelling at 20 km/h on a level road, the cyclist suddenly ceases to pedal, and in 10 seconds finds that his speed is changed to 16 km/h. What is the total force resisting motion? [U.E.I.]

11 What do you understand by acceleration? A force of 80 N causes a body to move 20 m in 4 seconds from rest. Calculate the uniform force required to move the body 60 m in 10 seconds if there is an initial velocity of 0·5 m/s. [N.C.T.E.C.]

12 A cage, together with its load, has a mass of 1·5 tonne, and is raised vertically by a rope. If the cage reaches a speed of 5 m/s after rising 30 m from rest with uniform acceleration, determine the tension in the rope during this period. Neglect the weight of the rope. [U.L.C.I.]

13 A car is travelling at a speed of 50 km/h when 400 m from a road intersection. The car is then uniformly accelerated and reaches 100 km/h at the road intersection. Find the time taken to cover the 400 m and the acceleration of the car. If the car weighs 12 kN and the frictional resistance is 200 N, find the tractive force exerted by the car. [U.L.C.I.]

14 A car weighing 8 kN starts from rest to climb a gradient of 1 in 10 against a frictional resistance of 180 N. If the tractive effort exerted by the car is 1800 N find (*a*) the acceleration of the car, (*b*) the time taken to reach 50 km/h, (*c*) the total distance travelled. What would be the acceleration on the level if the frictional resistance is unchanged? [U.L.C.I.]

15 A car travelling at 50 km/h commences to descend a gradient of 1 in 15 with the engine switched off. The car weighs 8 kN and the frictional resistance is 220 N. Find the force causing the car to accelerate, the acceleration of the car and the time taken to reach 100 km/h. Also find the total distance travelled by the car. [U.L.C.I.]

16 State Newton's Second Law of Motion.

A car, mass 1 tonne, and travelling at 50 km/h, is brought to rest in a distance of 150 m with a uniform retardation. Determine the value of the retarding force and the time taken in coming to rest. [N.C.T.E.C.]

17 A truck weighing 8 kN is being hauled up a plane, inclined at 20° to the horizontal, by means of a rope parallel to the plane. If the resistance to motion is equivalent to a force of 1·4 kN and the constant pull of the haulage rope is 4·5 kN, determine the acceleration of the truck and find the time required to travel 400 m along the incline. [N.C.T.E.C.]

Chapter Eight
Torque and Turning

You would discover very early in life that you could made a body turn about an axis if you applied a force to the body at some distance from the axis. You will have discussed in previous work the idea of the moment of a force about a point or the turning moment produced by force. The value of the turning moment is equal to the product of the magnitude of the force and the perpendicular distance from the point to the line of action of the force. We are going to give this turning moment a new name and call it **torque.**

Torque is a Latin word meaning a necklace which was intended to be twisted or bent around somebody's neck.

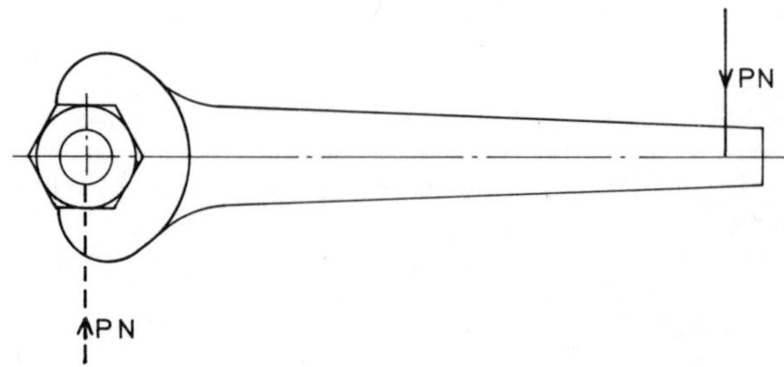

Figure 8.1 *Torque*

Immediately a force is applied at the end of the spanner there will be an equal and opposite force set up at the nut. These two forces form a 'couple' or a torque. Torques are always trying to produce turning.

A force at the end of a spanner applies a torque to the bolt or nut.

A pull in a belt passing over a pulley applies a torque to the shaft on which the pulley is mounted.

A pull in a cycle chain applies a torque to the sprocket wheel and hence to the rear road wheel.

The units of torque are, of course, the same as those for moments, i.e. newton metre (Nm).

Work done by Torque

Now we have seen that work is done when a force moves through a distance. In the same way work is done when a torque turns through an angle.

The arm OA, radius *r*m is fastened to a shaft at O. A force is applied at A perpendicular to OA, and causes the arm and the shaft to rotate. The magnitude of the force is F N (Fig. 8.2).

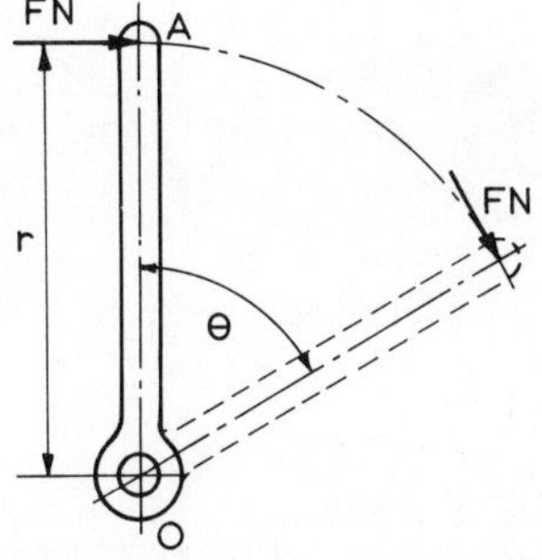

Figure 8.2 *Work done by a Torque*

The force F at the end of the lever OA causes the lever to rotate through angle θ rad. The work done by the torque is equal to the product of torque and angle in radians.

Let the arm move through an angle θ (measured in radians). Then the distance moved by A is given by the product of radius and angle turned through in radians.

A moves $r\theta$ metres

$$\text{Work done by the force} = \text{force} \times \text{distance}$$
$$= F \times r\theta$$
$$= Fr\theta \text{ newton metres}$$

But the product Fr is the moment of the force about O. It is equal to the torque applied to the shaft. Denote this torque by T Nm.

$$\therefore \quad \text{Work done by torque} = T\theta \text{ newton metres}$$

or

$$= T\theta \text{ joules.}$$

If the arm is rotating at a constant speed of ω rad/s it will move through an angle of ω radians in one second.

$$\text{Work done per second by torque} = T\omega \text{ joules per second.}$$

Now work done at the rate of 1 joule per second is equal to the unit of power of 1 watt.

$$\therefore \quad \text{Power being transmitted} = T\omega \text{ watts.}$$

Power transmitted (watts) = torque (newton metres) × angle turned through per second (ω).

Measurement of brake power for engine

The brake power of an engine is the power which is available at the output shaft for doing external work. It is usually measured by means of a brake; hence its name.

Fig. 8.3 shows an arrangement suitable for measuring the brake power of an engine. One end of a rope is hooked on to a spring balance suitably supported. The rope is wound round the flywheel or brake wheel, and weights are hung from the free end. The weights are arranged with respect to the direction of rotation, so that the engine tries to lift them.

Let W = dead weight, N
S = spring balance reading, N
r = radius of flywheel (measured to centre of rope), m
ω = angular velocity of engine, rad/s.

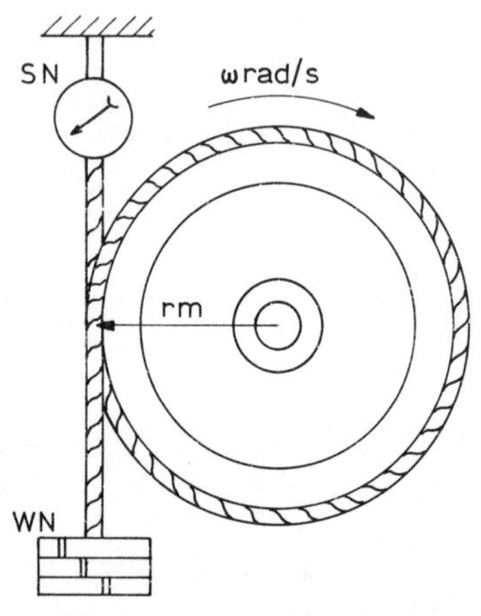

Figure 8.3 *Measurement of Brake Power of Engine*

The rope wrapped round the flywheel of the engine is fastened at one end to a spring balance and carries a weight at the other end. The resultant resisting torque on the engine shaft is $(W - S)r$ Nm. Knowing the speed of the engine, the power at the brake can be calculated.

$$\begin{aligned}\text{Resultant load at flywheel rim} &= (W - S)\text{N}\\ \text{Torque on shaft} &= \text{load} \times \text{radius}\\ &= (W - S)r\text{Nm}\\ \text{Power transmitted} &= \text{torque (Nm)} \times \text{angular velocity } \omega\\ &= (W - S)r\omega \text{ watts.}\end{aligned}$$

Example 8.1

How much work is done by a torque of 1000 Nm when applied to a shaft while it rotates through 50 revolutions?

Work done = torque × angle in rad
Angle turned through in radians

$$\begin{aligned}&= 50 \times 2\pi\\ &= 314{\cdot}2 \text{ rad}\end{aligned}$$

Work done = $1000 \times 314{\cdot}2$

$$\text{Work done} = 314\,200 \text{ joules.}$$

The work done by the torque is 314 200 joules.

Example 8.2

A shaft transmits 1000 kW at 600 rev/min. Calculate the torque in kNm on the shaft.

$$\begin{aligned} 600 \text{ rev/min} &= 20\pi \text{ rad/s} \\ 1000 \text{ kW} &= 10^6 \text{ joules per second} \\ \text{Power} &= \text{Torque} \times \text{angular velocity} \\ 10^6 &= T \times 20\pi \\ \text{Torque} &= 159\,000 \text{ Nm} \\ &= 159 \text{ kNm.} \end{aligned}$$

The torque on the shaft is 159 kNm.

Example 8.3

Determine the power of an engine from the following data obtained in a test:

$$\begin{aligned} \text{Average revolutions per minute} &= 220 \text{ rev/min} \\ &= 23{\cdot}1 \text{ rad/s} \\ \text{Load on brake} &= 400 \text{ N} \\ \text{Spring balance reading} &= 20 \text{ N} \\ \text{Brake diameter} &= 1{\cdot}5 \text{ m} \\ \text{Resultant load at brake radius} &= 400 - 20 \\ &= 380 \text{ N} \end{aligned}$$

$$\begin{aligned} \text{Braking torque} &= 380 \times \frac{1{\cdot}5}{2} \\ &= 285 \text{ Nm} \\ \text{Brake power} &= \text{torque} \times \text{angular velocity} \\ &= 285 \times 23{\cdot}1 \text{ J/s} \\ &= 6538 \text{ watts} \\ \text{Brake power} &= 6{\cdot}54 \text{ kW.} \end{aligned}$$

Brake power of engine is 6·54 kW.

EXERCISE 8

1 A torque of 30 Nm is applied to a shaft whilst it rotates through 10 revolutions. How much work is done by the torque?

2 A shaft transmits 40 kW while rotating at 100 rev/min. What is the torque on the shaft?

3 What power is transmitted by a shaft rotating at 500 rev/min if the applied torque is 2500 Nm?

4 A flexible coupling is designed to transmit 80 kW at 800 rev/min. If the speed is to be reduced to 500 rev/min, what will be the maximum power which the coupling will now transmit? (NOTE. The torque to which the coupling is subjected will be the same in both cases.)

5 Define torque.

In a motor-car having wheels 1 m in diameter, the speed ratio between the engine shaft and the rear axle is 5 to 1. If at a uniform speed of 54 km/h the resistance to motion is 1000 N, calculate the torque and power at the engine shaft, assuming 30 per cent. of the power to be lost between the engine and the wheels. [N.C.T.E.C.]

6 An hydraulic brake connected to an aero-engine on test applies a torque which is balanced by a load of 3000 N at the end of an arm 0·8 m long. If the speed of the engine is 2200 rev/min, calculate its power. [N.C.T.E.C.]

Chapter Nine
Potential and Kinetic Energy

First, just a reminder that by energy we mean the ability to do work.

Potential energy

The potential energy of a body is the energy which a body possesses as a result of its position. An example of this would be the case of a heavy block raised some distance above the ground. Work has had to be done in lifting the block, and this work is stored in it as potential energy. If the block is allowed to fall freely, then it can be used to do work, as for example in a pile driver: the block, on falling, is allowed to strike the pile which is driven into the ground to provide a retaining wall (Fig. 9.1). If the block

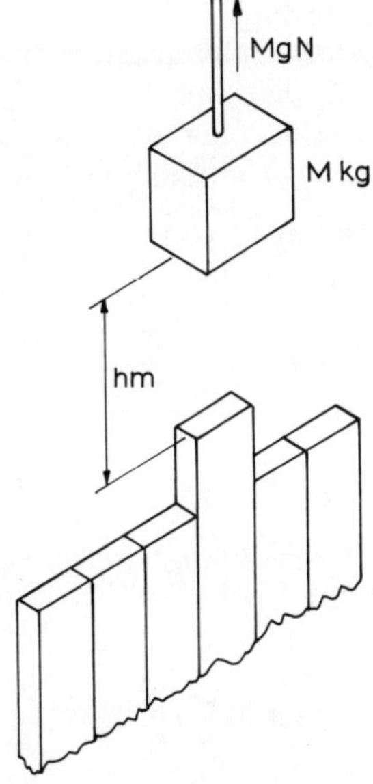

Figure 9.1 *Potential and Kinetic Energy*

The block of mass M is raised to a height h above the wooden plank. The block possesses potential energy. When the block is released, its potential energy is converted into kinetic energy. On colliding with the plank, the block gives up its kinetic energy by doing work in driving the plank farther into the ground.

were just allowed to rest on the pile, it would produce little effect, but by giving the block potential energy, i.e. by lifting the block over the pile, and then releasing it, it can be made much more effective in driving the pile into the ground.

Let M kg be the mass of the block, and let h m be the height to which it is raised.

Then the force required to lift a mass M kg at a uniform velocity is Mg newtons. Work done by this force is equal to force × distance, which is Mgh Nm. This work is stored in the body as potential energy.

$$\therefore \quad \text{Potential energy of body} = Mgh \text{ Nm}$$

where

$$M = \text{mass in kg}$$
$$h = \text{height in m}$$

Kinetic energy

The kinetic energy of a body is the energy which it possesses due to its velocity. In order to get the body moving, work has had to be done, and this work is stored in the body as kinetic energy. The body will give up this energy when it is brought to rest and in coming to rest can be made to do work. When we use a hammer to break a stone, we give the hammer kinetic energy by making it move quickly, and this energy is used in doing the necessary work to break up the stone.

Let us consider the motion of a body whose mass is M kg against which a force of F N is acting through a distance l metres.

$$\text{Work done by force} = \text{force} \times \text{distance}$$
$$= Fl \text{ Nm} \qquad (1)$$

$$\text{Acceleration of the body } f = \frac{\text{force}}{\text{mass}}$$

$$= \frac{F}{M} \qquad (2)$$

Velocity of the body after travelling l m from rest with this acceleration is given by

$$v^2 = 2fl \qquad (u = 0)$$

$$\therefore \quad l = \frac{v^2}{2f}$$

Substitute for f from (2)

$$= \frac{v^2M}{2F}$$

Substitute for l in (1)

$$\text{Work done} = Fl = \frac{Fv^2M}{2F} = \tfrac{1}{2}Mv^2 \text{ Nm}$$

The work done is therefore $\frac{1}{2}$ mass $\times$ velocity^2 Nm. This work has been stored in the body as kinetic energy.

$$\begin{aligned}\text{Kinetic energy of the body} &= \tfrac{1}{2}\text{ mass} \times \text{velocity}^2 \text{ Nm} \\ &= \tfrac{1}{2}Mv^2 \text{ Nm}\end{aligned}$$

NOTE M is in kg
v is in m/s

Conversion of potential and kinetic energy

We have seen that when a body of mass M kg is lifted to a height of h m, its potential energy is Mgh Nm. If the body is allowed to fall, its potential energy is reduced, since the height is reduced. On the other hand, the kinetic energy of the body is increased, since its velocity is increased. Now, if no external work is done, the total energy which the body possesses remains constant.

$$\text{Potential energy} + \text{kinetic energy} = \text{constant.}$$

When the body is at rest at height h, the only energy which it possesses is potential.

$$\begin{aligned}\text{Total energy before falling} &= \text{potential} + \text{kinetic} \\ &= Mgh + 0 \\ &= Mgh \text{ Nm} \qquad (1)\end{aligned}$$

Since no work is done by the body whilst falling through h m, the total energy on reaching the ground must equal Mgh. At this instant the only energy possessed by the body is kinetic.

$$\begin{aligned}\text{Total energy on reaching ground} &= \text{potential} + \text{kinetic} \\ &= 0 + \tfrac{1}{2}\text{ mass} \times \text{velocity}^2 \\ &= \tfrac{1}{2}Mv^2 \qquad (2)\end{aligned}$$

Equating equations (1) and (2), we have:

$$\begin{aligned}Mgh &= \tfrac{1}{2}Mv^2 \\ v^2 &= 2gh\end{aligned}$$

Velocity of body on reaching ground

$$v = \sqrt{2gh}$$

which is the same expression as we obtained previously.

For any intermediate part of the fall, the body will have both potential and kinetic energies. The sum of these two must still equal Mgh, provided no work is done by the body in falling.

Example 9.1

What is the kinetic energy of a motor-car of mass 2 tonnes when travelling at a speed of 54 km/h? If the speed drops to 9 km/h in 4 seconds, what is the change of energy per second?

$$\text{Mass of car} = 2 \times 1000 \text{ kg}$$
$$= 2000 \text{ kg}$$
$$\text{Velocity} = \frac{54 \times 1000}{60 \times 60}$$
$$= 15 \text{ m/s}$$
$$\text{Kinetic energy} = \tfrac{1}{2} \text{ mass} \times \text{velocity}^2$$
$$= \tfrac{1}{2} \times 2000 \times 15^2$$
$$\text{Kinetic energy} = 225\,000 \text{ Nm at } 54 \text{ km/h}$$
$$\text{Velocity at } 9 \text{ km/h} = \frac{9 \times 1000}{60 \times 60}$$
$$= 2{\cdot}5 \text{ m/s}$$
$$\text{Kinetic energy} = \tfrac{1}{2} \times 2000 \times 2{\cdot}5^2$$
$$= 6250 \text{ Nm}$$
$$\text{Change of energy} = 225\,000 - 6250$$
$$= 218\,750$$
$$\text{Change of energy per second} = \frac{218\,750}{4}$$
$$= 54\,687{\cdot}5 \text{ Nm/s.}$$

Change of energy per second is 54·7 kNm/s.

NOTE. This figure will be useful in the design of suitable brakes for stopping the car. The change in the kinetic energy of the car will appear as heat energy at the brakes.

Example 9.2

A flywheel has a mass of 4000 kg which may be considered as concentrated at a radius of 2 m. Calculate its kinetic energy at 200 rev/min. If the wheel slows down to 198 rev/min, how much energy is given up by the flywheel?

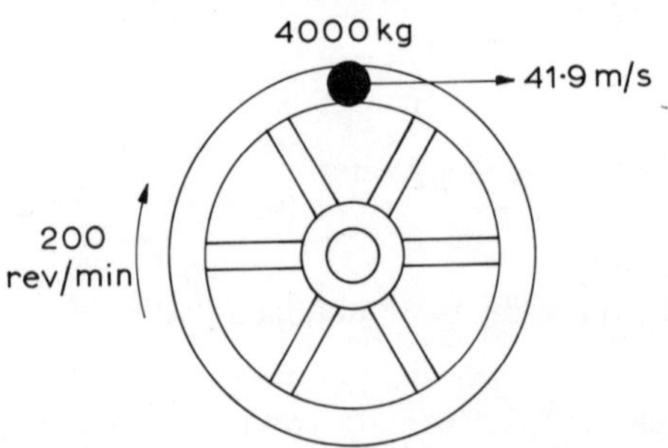

Figure 9.2 *Kinetic Energy of Rotation*

The whole mass of the flywheel is considered as acting at the rim. The kinetic energy of the wheel is, then, the same as the kinetic energy of the concentrated mass moving with a linear velocity of 41·9 m/s.

$$\text{Mass of flywheel} = 4000 \text{ kg}$$

Linear velocity of point on flywheel at 2 m radius

$$= \omega r$$
$$= \tfrac{200}{60} \times 2\pi \times 2$$
$$= 41{\cdot}9 \text{ m/s}$$

$$\text{Kinetic energy} = \tfrac{1}{2} \text{ mass} \times \text{velocity}^2$$
$$= \tfrac{1}{2} \times 4000 \times 41{\cdot}9^2$$
$$= 3\,511\,000 \text{ Nm}$$

$$\text{Kinetic energy at 200 rev/min} = 3511 \text{ kNm}$$

$$\text{Linear velocity at 198 rev/min} = \tfrac{198}{60} \times 2\pi \times 2$$
$$= 198 \times \frac{\pi}{15} \text{ m/s}$$

$$\text{Kinetic energy at 198 rev/min} = \tfrac{1}{2} \times 4000 \times \left[198 \frac{\pi}{15}\right]^2 \text{ Nm}$$

$$\text{Kinetic energy at 200 rev/min} = \tfrac{1}{2} \times 4000 \times \left[200 \frac{\pi}{15}\right]^2 \text{ Nm}$$

$$\text{Change in energy} = \tfrac{1}{2} \times 4000 \left(\frac{\pi}{15}\right)^2 [200^2 - 198^2] \text{ Nm}$$

Applying the 'difference of two squares' rule, we get:

$$\text{Change in energy} = \tfrac{1}{2} \times 4000 \times \frac{\pi^2}{225} (200 + 198)(200 - 198)$$
$$= \tfrac{1}{2} \times 4000 \times \frac{\pi^2}{225} \times 398 \times 2$$
$$= 174\ 600 \text{ Nm}$$

The kinetic energy at 200 rev/min is 3511 kNm and the change in energy is 174·6 kNm.

NOTE. This energy is given up by the flywheel because the engine will not have produced sufficient energy to overcome the external resistance at that particular instant.

Example 9.3

A pump whose efficiency is 70 per cent. lifts 2000 litres of water per minute through a height of 200 m. Resistances in the pipe are equivalent to an extra head of 40 m. Calculate the work done by the pump per minute. What would be the power of the motor required to drive the pump? 1 litre of water has a mass of 1 kg.

Mass of water to be lifted per min

$$= 2000 \text{ kg}$$

Force required to lift this mass $= 2000\,g = 19\,620$ N

Equivalent height to which water is raised

$$= 200 + 40$$
$$= 240 \text{ m}$$

Work done in lifting water per min through 240 m
$= 19\,620 \times 240$
$= 4\,708\,800$ Nm or joules

Since only 70 per cent. or 0·7 of the energy supplied to the pump is used in pumping the water

$$\text{Energy supplied to pump} = \frac{4\,708\,800}{0{\cdot}7}$$
$= 6\,726\,800$ joules per min
$= 112\,113$ joules per second
1 joule per second $= 1$ watt

Power of motor supplying this energy to the pump
$= 112\,113$ watts
$= 112{\cdot}1$ kW

Work done by pump per minute is 4 708 400 joules.
Power of motor is 112·1 kW.

EXERCISE 9

Remember that the weight of a body mass M *kg is always* Mg *newtons.*

1 A motor-car of 1 tonne mass is travelling at 54 km/h. What is the kinetic energy of the car? How much kinetic energy is lost when the car reduces speed to 9 km/h?

2 A train mass 200 tonnes travelling at 45 km/h is brought to rest in a distance of 2·5 km by a constant braking force. What is the value of this force? How much work is done by this force in bringing the train to rest? What is the change in the kinetic energy of the train?

3 A mass of 20 kg is raised by a constant force of 250 N. What is the kinetic energy of the mass after moving through 10 m from rest?

4 A flywheel of 3 tonnes mass has a mean diameter of 2 m. If the driving torque is shut off when the speed of the wheel is 135 rev/min and the energy used in overcoming friction in the bearings is 1500 Nm per revolution, how many revolutions will it make before its speed has fallen to 75 rev/min?

5 (*a*) A flywheel has a mass of 2000 kg and this mass may be assumed concentrated at a radius of 1·2 m. Calculate its kinetic energy at 150 rev/min.
(*b*) If the wheel slows down from 150 to 148 rev/min, find the energy abstracted from the wheel. [U.E.I.]

6 The rim of a flywheel has a mass of 6 tonnes and it has a mean diameter of 3 m.
(*a*) Find the kinetic energy of the rim when rotating at 120 rev/min.
(*b*) If the speed drops to 50 rev/min, find the change in kinetic energy. Give your answer in kilojoules. [U.E.I.]

7 Distinguish between kinetic and potential energy.
A mass of 10 kg falls freely from rest through a vertical height of 6 m when its velocity is decreased to 6 m/s by imparting energy to a machine. Calculate the energy in joules given to the machine. [U.L.C.I.]

8 A bus, mass 3 tonnes, reduced its speed from 18 km/h to 4·5 km/h. Determine the loss in kinetic energy of the bus in kilojoules and also the retarding force in newtons if the retardation took place in 12 seconds.

9 Distinguish clearly between potential and kinetic energy.

A hammer, mass 6·5 kg, is held in a horizontal position and then released so that it swings in a vertical circle of 1 metre radius. What is the kinetic energy and the speed of the hammer at the lowest point of the swing?

If at this point it breaks a piece of metal and so parts with 30 joules of energy, to what height will it rise on the other side? [U.L.C.I.]

10 Define (*a*) potential energy and (*b*) kinetic energy.

A drop hammer is raised, between guides, to a height of 3 m above its striking point—and released. At the instant of striking, it is found to have a velocity of 6 m/s. If the hammer has a mass of 1 tonne, how much energy has been lost during its fall? If the resisting force may be assumed constant, what is its value? [N.C.T.E.C.]

11 Define kinetic energy and potential energy.

An object mass 10 kg falls freely through a distance of 60 m on to the roof of a building. If the speed of the object immediately after penetrating the roof is 10 m/s, how much energy was used in the penetration? [N.C.T.E.C.]

12 Define (*a*) work (*b*) power and (*c*) energy.

A pump is required to lift 1800 litres of water per minute through a height of 100 m. If the resistance in the pipes is equivalent to an additional head of 10 m, calculate the work done by the pump per minute. If the efficiency of the pump is 70 per cent., calculate the power of the motor required to drive it. [N.C.T.E.C.]

13 A loaded lift of total mass, including support ropes, of 2000 kg is raised 10 m above ground level. The mass of the supporting ropes is 15 kg per metre length.

Determine: (*a*) the total work done during the 10 m lift, (*b*) the work done during the first 3 m of lift, and (*c*) the instantaneous power required if the speed of the lift at a point 3 m from the ground is 18 km/h. [N.C.T.E.C.]

14 A flywheel has its speed uniformly increased from 40 rev/min to 280 rev/min in 3 min. Find the acceleration in rad/s^2, and the total number of revolutions made in 3 min. If the torque applied to the flywheel during the 3 min is 6·5 Nm, find the increase in kinetic energy of the flywheel. If the radius of the flywheel is 0·6 m, what is the linear speed of a point on the rim of the flywheel when it is revolving at 280 rev/min? [U.L.C.I.]

15 A flywheel revolving at 300 rev/min has its speed uniformly reduced in 2 minutes. If the average speed during the 2 minutes is 210 rev/min, find the final speed. Find the angular retardation in rad/s and the total number of revolutions made by the flywheel in the 2 minutes. If the flywheel loses 45 kJ of energy during the reduction of speed, find the braking torque which must be applied. [U.L.C.I.]

16 Define (i) potential energy, (ii) kinetic energy.

A loaded lift mass 2000 kg is raised by a constant force of 20 kN. Determine: (*a*) its kinetic energy after moving 3 m from rest, (*b*) its velocity at this point. [N.C.T.E.C.]

Chapter Ten
Motion in a Circular Path

Change in velocity

When discussing the question of velocity and acceleration, we mentioned that, since 'velocity' involved both magnitude and direction, we could have a change of velocity if either the magnitude or the direction of the velocity were changed. Thus, a car travelling on a straight road with a velocity at successive intervals of 10, 15, 20, 25, 30 km/h is accelerating. What is perhaps not quite so obvious is that if the car were travelling on a curved path with a constant speed (you remember that speed is only the number part—no direction involved in speed), so that after successive intervals its direction is given by N, N 15° E, N 30° E, N 45° E, then this car is also accelerating. It is this kind of acceleration which we are to review now.

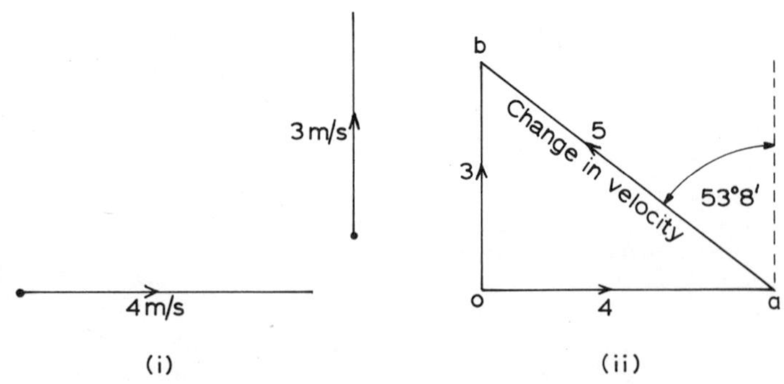

Figure 10.1 *Acceleration due to Change in Direction*

If a body changes its direction of motion, it is accelerating even though the magnitude of the velocity remains constant. The change in velocity is the vector difference of the two velocities.

First of all, let us see how we can obtain the change in velocity from a velocity triangle. A body is travelling at 4 m/s in a direction due east. Ten seconds later it is travelling at 3 m/s in a direction due north. The two velocities can be represented by the vectors as in Fig. 10.1(i). Now draw the velocity triangle.

oa represents 4 m/s due east: the initial velocity.
ob represents 3 m/s due north: the final velocity.
From the vector triangle we know that:

$$\boldsymbol{oa} + \boldsymbol{ab} = \boldsymbol{ob}$$
$$\boldsymbol{ob} - \boldsymbol{oa} = \boldsymbol{ab}$$

So that ***ab*** represents the difference between the final velocity and the initial velocity.

ab is therefore the change in the velocity, and equals 5 m/s in a direction 53° 8′ west of north. This change has taken place in 10 seconds, so that the change in velocity per second is $\frac{5}{10}$ or 0·5 m/s^2, which is the acceleration. Of course, this must be an average rate, since we are not told whether the change in the velocity was gradual or instantaneous. In any case, we were only interested in the change which is given by the line ***ab***. Now there is no doubt that this is not easy to understand, but we must feel fairly well satisfied about it before we can understand the rest of the story.

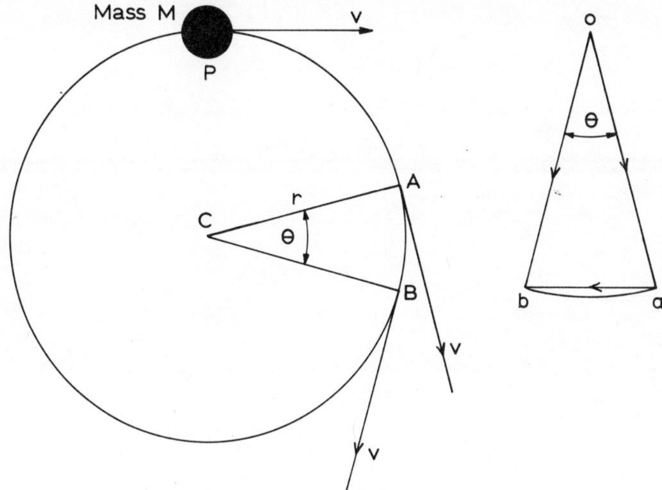

Figure 10.2 *Motion in a Circle*

The direction of the velocity of P is constantly changing. It is therefore subjected to a special acceleration known as centripetal acceleration.

Circular motion

P is a body whose mass is M and which is travelling in a circular path of radius r with centre C (Fig. 10.2). Its velocity will always be tangential to the path and will therefore be changing in direction. We shall, however, consider that the magnitude of the velocity remains constant and equal to v. P passes through two points A and B during a time interval of t seconds (and we must think of t as being a very small time interval, shall we say $\frac{1}{100}$ second, or, if you like, much smaller still). Let the angle between the lines CA and CB be θ rad (and since we think of t as being very small, A and B will be very near together and θ will be a very small angle).

Now draw the velocity triangle. ***oa*** represents the velocity at A, ***ob*** represents the velocity at B, and ***ab*** represents the difference between these velocities. Since ***oa*** is perpendicular to CA and ***ob*** is perpendicular to CB, the angle between ***oa*** and ***ob*** will be the same as between CA and CB, which was that very small angle θ. Now we want the length of *ab*. If we draw an arc *ab* from centre *o*, then, provided θ is small, we shall not be able to tell the difference between the length of the arc *ab* and the length of the straight lines ***ab***.

$$\text{Arc length } ab = \text{radius} \times \text{angle in rad}$$
$$= \boldsymbol{oa} \times \theta$$
$$\therefore \quad \boldsymbol{ab} = v\theta.$$

$$\text{Change in velocity} = \boldsymbol{ob} - \boldsymbol{oa}$$
$$= \boldsymbol{ab}$$
$$= v\theta \text{ in } t \text{ seconds.}$$

$$\text{Change in velocity per second} = \frac{v\theta}{t} \qquad (1)$$

But $$\frac{\theta}{t} = \text{angular velocity of the line CA (rad/s)}$$
$$= \omega$$
$$= \frac{v}{r}$$

$$\text{Change in velocity per second} = v.\frac{v}{r} = \frac{v^2}{r}, \text{ from (1)}$$

Since $$v = \omega r, \qquad \text{we can write } \frac{v^2}{r} = \frac{\omega^2 r^2}{r} = \omega^2 r$$

$$\text{Acceleration of P} = \frac{v^2}{r} \text{ or } \omega^2 r.$$

Direction of acceleration

Since θ is very small, the sum of the two angles $\angle bao$ and $\angle abo$ must be nearly 180°, and since the triangle is isosceles, each angle must be practically 90°. The smaller we make θ the more correct does this statement become.

The direction of the acceleration is perpendicular to the direction of the velocity. It is therefore acting along the radius *towards the centre of the circle C*. For this reason it is known as a centripetal acceleration.

So, summarising, we can say that when a body is moving with velocity v round a path of radius r it has an acceleration of $\frac{v^2}{r}$ towards the centre of the circular path, this acceleration being known as the *centripetal* acceleration.

Centripetal and centrifugal force

Now, we never get an acceleration unless we have a force producing it. If P has an acceleration towards the centre of the circular path, there must be a force acting towards this centre producing this acceleration.

The mass of P is M

$$\text{Force} = \text{mass} \times \text{acceleration}$$

$$= M \times \frac{v^2}{r}$$

or alternatively

$$= M\omega^2 r$$

This force is known as the *centripetal* force, *which must always be applied to make a body move in a circular path.*

If the mass is measured in kilograms, the velocity in metres per second and the radius in metres, then the centripal force will be measured in newtons.

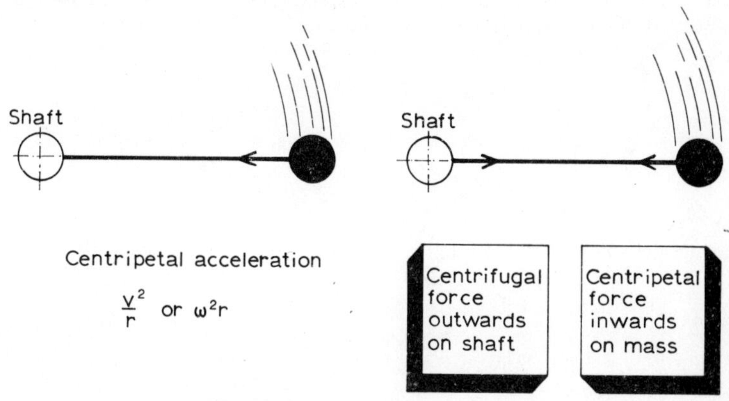

Figure 10.3 *Centripetal and Centrifugal*

As the mass is whirled around the shaft, it experiences a centripetal acceleration towards the centre of the shaft. There must therefore be a force known as a centripetal force pulling the mass towards the centre of rotation. The shaft experiences an equal and opposite reaction known as a centrifugal force.

When you make a stone move in a circular path at the end of a string, the string is all the time *pulling the stone into a circular path* with your hand as centre. Otherwise the stone would move off in a straight line.

When a motor-cyclist travels round the 'wall of death,' the vertical boardings are all the time *pushing the cyclist into a circular path.*

But while the string is pulling the stone towards the centre, it is also true that there is an equal and opposite reaction on your hand at the centre. This opposite reaction is known as the *centrifugal* force. It is equal in magnitude to the centripetal force, and only exists as an equal and opposite reaction to the centripetal force.

Example 10.1

A point P is moving around a circular path with a velocity of 10 m/s. If the radius of the path is 4 m, what is the magnitude and direction of the acceleration of P?

The centripetal acceleration of P is $\dfrac{v^2}{r}$

$$= \frac{10 \times 10}{4}$$

$$\text{Acceleration} = 25 \text{ m/s}^2.$$

Acceleration is 25 m/s² in direction towards the centre of the circular path.

Example 10.2

The crankpin of an engine is 100 mm from the crankshaft, which is rotating uniformly at 200 rev/min. What is the acceleration of the crankpin?

The centripetal acceleration of the crankpin is $\omega^2 r$.

$$\begin{aligned} \omega &= \tfrac{200}{60} \times 2\pi \text{ rad/s} \\ &= 20{\cdot}95 \text{ rad/s} \\ r &= 100 \text{ mm} = 0{\cdot}1 \text{ m.} \end{aligned}$$

$$\begin{aligned} \text{Centripetal acceleration} &= 20{\cdot}95^2 \times 0{\cdot}1 \\ &= 43{\cdot}9 \text{ m/s}^2. \end{aligned}$$

The acceleration of the crankpin is 43·9 m/s² towards the crankshaft.

Example 10.3

In a machine, a horizontal arm 0·5 m long is driven by a vertical shaft and makes 600 rev/min in a horizontal plane. A block of mass 5 kg is attached to the free end of the arm.

Determine:

(*a*) the force on the block;
(*b*) the force on the vertical shaft due to the rotation of the block.

(*a*)

$$\text{Mass of block} = 5 \text{ kg}$$

$$\begin{aligned} \text{Centripetal acceleration of block} &= \omega^2 r \\ &= \left(\frac{600 \times 2\pi}{60}\right)^2 \times 0{\cdot}5 \\ &= 1974 \text{ m/s}^2. \end{aligned}$$

$$\text{Force to produce this acceleration} = \text{mass} \times \text{acceleration}$$
$$= 5 \times 1974$$
$$= 9870 \text{ newtons.}$$

There is a centripetal force of 9870 N pulling the block towards the centre of circle.

(*b*) There is a force of the same magnitude 9870 N, but opposite in direction, which the arm exerts on the vertical shaft. This force will tend to cause the vertical shaft to deflect towards the block.

This force is known as the centrifugal force.

Example 10.4

A body of mass 200 kg rotates on frictionless rails inside a vertical circle of 10 m radius. Determine the force exerted by the rails on the body:

(*a*) at the top, and
(*b*) at the bottom of the circle,

if the linear velocity of the body is 50 m/s.

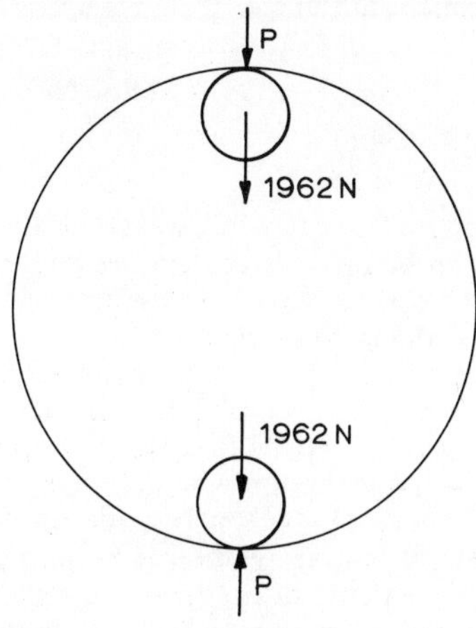

Figure 10.4

$$\text{Mass of the body} = 200 \text{ kg}$$
$$\text{Weight of body} = Mg$$
$$= 200 \times 9{\cdot}81 \text{ newtons}$$
$$= 1962 \text{ newtons}$$

$$\text{Centripetal acceleration of body} = \frac{v^2}{r} = \frac{50^2}{10}$$
$$= 250 \text{ m/s}^2$$

Centripetal force producing this acceleration

$$= \text{mass} \times \text{acceleration}$$
$$= 200 \times 250 \text{ newtons}$$
$$= 50 \text{ kN}.$$

There must always be a force of 50 kN pushing the mass towards the centre of the circle.

(*a*) Forces acting on body at the top of the circle:

(1) Force exerted by rails on body, P N down
(2) Weight of body, 1962 N down

Resultant force towards the centre $= P + 1962$ N
This is the necessary centripetal force of 50 000 N

$$\therefore \quad P + 1962 = 50\,000$$
$$P = 48\,038 \text{ N}.$$

(*b*) Forces acting on body at bottom of the circle:

(1) Force exerted by rails on body P N up
(2) Weight of body 1962 N down

Resultant force towards centre $= P - 1962$ N
This is the necessary centripetal force

$$\therefore \quad P - 1962 = 50\,000$$
$$P = 51\,962 \text{ N}.$$

Rail force at the top $= 48\,038$ N
Rail force at the bottom $= 51\,962$ N.

NOTE. 1. The rail force at the top is always less than that at the bottom. If the rail force at the top becomes zero, then the body is on the point of leaving the rails. This fact helps to determine the least linear velocity which the body can have to retain contact with the rails.
2. It has been assumed in the above solution that the linear velocity of the body is maintained at a constant value. This would mean that some type of drive would have to be applied. If the body were moving independently of ar additional drive, its velocity at the top would be less than that at the bottom The total energy of the body at the top is made up of potential energy anc kinetic energy, whilst at the bottom the body has no potential energy but onl kinetic energy. Hence, since the total energy at the top and at the bottom mus be the same, the kinetic energy at the top will be less than that at the bottom This means that the velocity at the top will be less than the velocity at th bottom.

Balancing of rotating masses

Single Rotating Mass. If a mass of M at a radius of R is fastened to a shaf rotating at ω, there is a centrifugal force of $M\omega^2R$ acting outwards on th shaft, all of which produces an out of balance effect on the shaft, whic would be very detrimental to the efficient running of the machine, producin excessive wear on the bearings, and setting up vibrations.

To balance the shaft it would be necessary to place a balance mas immediately opposite the original mass M, so that the centrifugal force o

Figure 10.5 *Balance Masses*

the shaft due to the balance mass would be equal and opposite to that set up by the mass M.

If B is the mass of the balance mass, which is placed at radius r, then the centrifugal force would be $B\omega^2 r$ so that for balance

$$B\omega^2 r = M\omega^2 R$$

or

$$Br = MR$$

So that the product of the mass and the radius must be the same for the balance mass and the original out of balance mass. The product Br or MR is referred to as a mass moment.

Several Rotating Masses. If there are several rotating masses connected to the shaft at different radii, but all in one plane perpendicular to that of the shaft, then each sets up centrifugal forces on the shaft. It will be necessary to supply only one balance mass whose magnitude can be determined by means of a force diagram. Since all the masses are connected to the same shaft, they have a common angular velocity ω. We need not calculate the magnitude of the centrifugal force, but deal only with mass moment products.

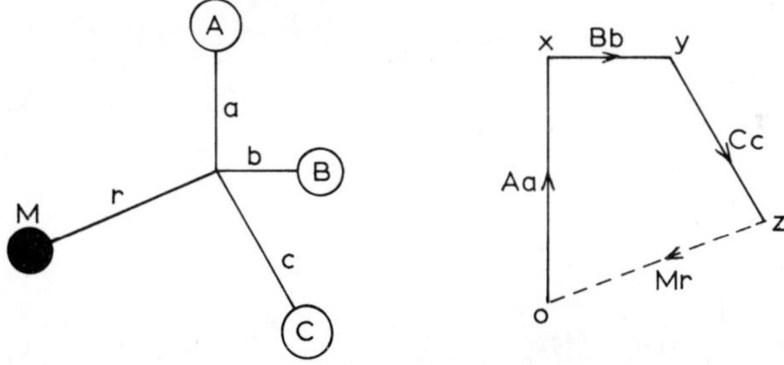

Figure 10.6 *Balancing Several Masses by a Single Balance Mass*

Three masses, A, B and C are fastened to the shaft at radii a, b and c respectively, their angular positions being as shown in Fig. 10.6. We are to determine the size of a balance mass M to be placed at radius r to balance the system. The method is as follows:

(1) Calculate the mass moment for each mass.
(2) Draw the vector diagram for these mass moments:

Commencing at o, draw ox to represent Aa
xy to represent Bb
yz to represent Cc

(3) The closing line *zo* (from *z* to *o*) indicates the magnitude and direction of the balancing mass moment *Mr*.
(4) Measure *zo* to scale and, knowing *r*, calculate the value of *M*.

Note the line *oz* (from *o* to *z*) gives the magnitude and direction of 'out of balance' mass moment. To convert this into a force we must multiply by ω^2.

Example 10.5

A circular plate revolving about its centre at 150 rev/min has a mass of 10 kg attached at a radius of 40 mm and a mass of 8 kg at a radius of 60 mm. The angle between the masses is 90°. Determine the magnitude and direction of the out of balance force on the shaft.

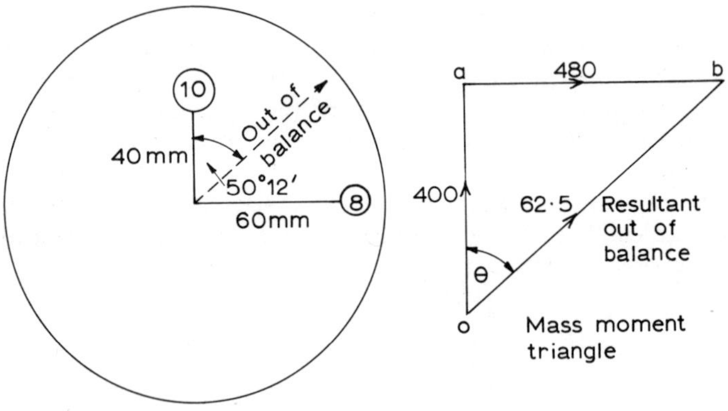

Figure 10.7

The position of the masses is shown in Fig. 10.7.

Mass moment of 10 kg = 10 × 40 = 400 kg mm
Mass moment of 8 kg = 8 × 60 = 480 kg mm
Draw *oa* in direction of the 40 mm arm to represent 400 kg mm
Draw *ab* in direction of the 60 mm arm to represent 480 kg mm

The resultant out of balance mass moment is given by the line *ob* which represents 625 kg mm.

To convert this into a force, we must multiply by ω^2. Also the 625 kg mm must be converted into kg m, to give the value of the force in newtons, since by definition, the newton is associated with an acceleration of 1 m/s^2.

$$\therefore \quad \text{Out of balance force} = \frac{625}{1000} \times \left(\frac{150}{60} \times 2\pi\right)^2$$

$$= 154{\cdot}2 \text{ newtons.}$$

Direction of out of balance force:

$$\tan\theta = \tfrac{48}{40} = 1{\cdot}2$$
$$\theta = 50^\circ\ 12'$$

The out of balance force of 154·2 newtons makes an angle of 50° 12′ with the 40 mm arm as shown.

Example 10.6

Three masses are attached to a rotating shaft: 16 kg at 5 cm radius, 20 kg at 6 cm radius and 10 kg at 9 cm radius. The angle between the 16 kg and the 20 kg masses is 120°, and the angle between the 20 kg and the 10 kg masses is 90°. Determine the magnitude and position of a balance mass to be placed at a radius of 4 cm. All the masses are in the same plane.

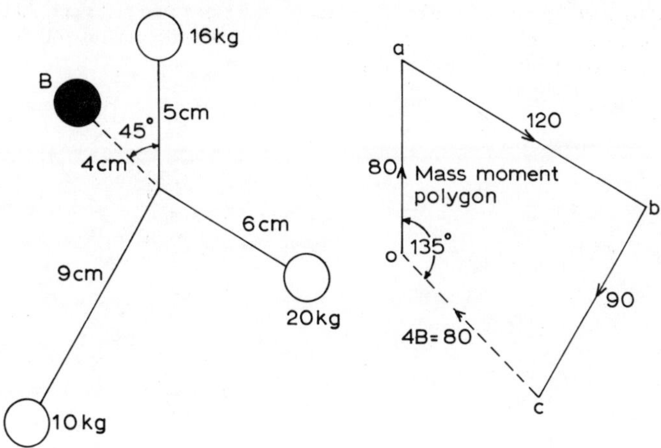

Figure 10.8

1 Construct the following table:

Mass kg	Radius cm	Mass moment kg cm
16	5	80
20	6	120
10	9	90
B	4	4*B*

2 Draw the mass moment polygon, using the values from the last column.

Start at *o* (Fig. 10.8), draw *oa* to represent 80 cm in direction of 5 cm arm.
Draw *ab* to represent 120 kg cm in direction of 6 cm arm.
Draw *bc* to represent 90 kg cm in direction of 9 cm arm.

Then *co* represents the magnitude and direction of the balancing mass moment $4B$.

co measures 80 kg cm units.

$$\therefore \quad 4B = 80$$
$$B = 20 \text{ kg}$$

Angle *aoc* measures 135°.

The balance-mass arm is drawn parallel to the line *co*, thus making 45° with the direction of the 5 cm arm.

The balance mass has a magnitude of 20 kg, and is set at 4 cm radius in direction making 45° with the 5 cm radius as shown.

Example 10.7

Three masses are attached to a shaft as follows: 20 kg at 3 cm radius, 30 kg at 4 cm radius and 18 kg at 5 cm radius. The masses are to be arranged so that the shaft is in balance. Determine the angular position of the masses relative to the 20 kg mass. All the masses are in the same plane.

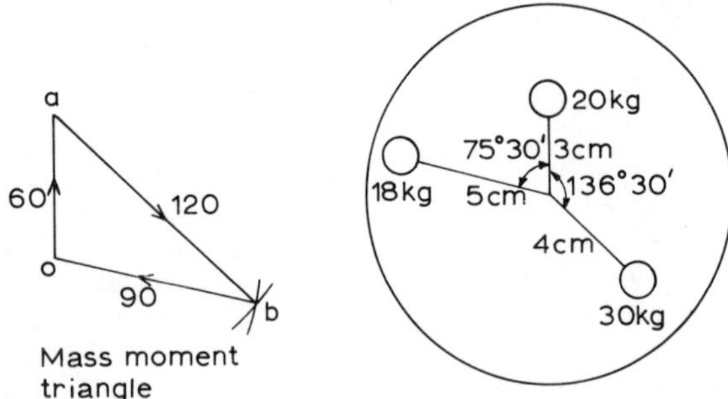

Figure 10.9

1 Construct the following table:

Mass kg	Radius cm	Mass moment kg cm
20	3	60
30	4	120
18	5	90

2 Since the masses are in equilibrium there will be no out of balance mass moment, and therefore the mass moment diagram will be a closed triangle whose sides are proportional to the values in the last column.

Draw *oa* to represent 60 kg cm in any convenient direction.
With *a* as centre and radius equal to 120 kg cm, draw arc of circle.
With *o* as centre and radius equal to 90 kg cm, draw arc of circle intersecting first arc at *b*.

Join *ab* and *bo*. Then the triangle *oab* is the required mass moment triangle which gives the relative directions of the three masses.

3 Draw the 4 cm arm parallel to *ab*.
Draw the 5 cm arm parallel to *bo*.
Measure the angle between the arms

Angle between 20 kg and 30 kg is 136° 30′ clockwise.
Angle between 20 kg and 18 kg is 75° 30′ anticlockwise.

NOTE. *To avoid confusion in these problems, it is always advisable to indicate the angular positions of the masses on a diagram.*

EXERCISE 10

1 Determine the acceleration of a point moving with a constant speed of 50 m/s round a circular path of radius 3 m.

2 A flywheel, 2 m in diameter, is rotating at 400 rev/min. What is the acceleration of a point on the rim of the wheel?

3 The centripetal acceleration of a crankpin is 300 m/s^2. If the crank radius is 1 m, at what speed is the engine running?

4 A motor-cyclist rides round a 'wall of death' at a speed of 60 km/h. If the wall has a diameter of 30 m, calculate the force between the wall and the motor-cycle wheels. The mass of the cycle and rider can be taken as 200 kg.

5 The maximum safe load which a piece of string can carry is 400 N. The string is used to swing a 1 kg mass in a horizontal circle at a radius of 1 m. What is the greatest velocity which the mass can have without breaking the string?

6 A gear wheel, mass 300 kg, has its centre of gravity 0·5 cm from the axis of rotation. Calculate the magnitude of the unbalanced force on the shaft when rotating at 300 rev/min.

7 A train whose mass is 300 tonnes is travelling at 90 km/h round a circular track whose radius is 6000 m. What will be the thrust exerted by the rails on the wheel flanges?

8 A 'Big Dipper' truck has a loaded weight of 20 kN. If the truck moves over the crest of a hump of radius 30 m at a speed of 54 km/h, determine the force between the truck and the rails.

9 Two masses, 4·5 kg and 9 kg, are attached in the same plane to a shaft at radii of 120 mm and 200 mm respectively. The angle between the masses is 120°. Determine the angular position and the magnitude of a balance weight which, when placed at a radius of 150 mm, will produce equilibrium.

10 Three masses are attached to a circular faceplate:

10 kg at 200 mm radius
7 kg at 150 mm radius
14 kg at 100 mm radius.

Determine the angular position of these masses so that there shall be no unbalanced force on the shaft.

11 A casting and two balance masses are bolted to the faceplate of a lathe so that the system is in perfect balance. The casting mass is 14 kg and its centre of gravity is 250 mm from the axis of rotation. The two balance masses are 7 kg and 10 kg, and are placed at radii of 300 mm and 280 mm respectively from the axis of rotation. Determine the angular positions of the casting and the masses, showing the arrangement on a diagram with the values of the angles inserted. [U.L.C.I.]

12 A circular plate which is revolving about its centre at 200 rev/min has a mass of 10 kg attached to it at a radius of 200 mm and another of 7 kg at a radius of 300 mm. The angle between these two radii is 120°. Determine the out of balance force, and state its direction relative to the given radii. [U.L.C.I.]

13 A body whose mass is 1000 kg rotates on frictionless rails inside a vertical circle of 10 m radius. Find the least speed (linear) of the body in m/s which will enable it to retain contact with the rails when at the highest point in the circle. [U.E.I.]

14 A mass, 70 kg placed at a radius of 300 mm rotates at a speed of 100 rev/min.
(*a*) Calculate the centrifugal force due to this revolving mass.
(*b*) What mass must be placed at a radius of 400 mm and directly opposite the given mass to balance the centrifugal force set up during rotation? [U.E.I.]

15 Two masses A and B revolve at 240 rev/min in the same plane. B is on a radius 120° anticlockwise from A. A, mass 5 kg, is 250 mm from the centre of rotation, and B, mass 10 kg, is 300 mm from the centre of rotation. Find the magnitude of the unbalanced centrifugal force. Find also the position and magnitude of the mass which will give dynamic balance if it is 250 mm from the centre of rotation. [U.L.C.I.]

16 A 10 kg mass is whirled in a vertical plane at 80 rev/min at the end of a wire 1·2 m in length. Find the centrifugal tension in the wire, and the maximum and minimum tensions in the wire. What is the minimum speed in rev/min at which the mass can be whirled to keep the wire taut? [U.L.C.I.]

Chapter Eleven
Machines

Mechanical advantage and velocity ratio

Very early in his history, man found that he was called upon to carry out tasks which were made difficult, if not impossible, by his own physical limitations. Even the strongest man discovered that there was a limit to the mass which he could lift, and to the load which he could drag over the ground. He therefore tried to devise some means to do the required work without using more effort than was necessary. For example, he found that a heavy mass could be lifted easily if a lever was used. He found that a large block of stone could be moved more easily over the ground if a number of rollers were placed under the block and the block pushed over the rollers instead of being dragged over the ground. He found that very heavy stone blocks could be brought more easily into position, in such buildings as the Pyramids, if they were brought gradually up a slope or inclined plane. It would have been impossible to lift some of these stones into position since they had no cranes or even 'sky-hooks' available. They had, however, a considerable amount of slave labour available, and this labour could be employed in producing a force parallel to the plane in order to pull the blocks into position. In all these instances, one fundamental point is to be found: a large resistance or load was overcome by means of a relatively small force or effort. Any arrangement which will produce this result is called a machine. The sturdy branches of trees used as levers, cylindrical tree trunks used as rollers, or the long slope of rubble and brick and stone were just as much machines as the elaborate cranes or the impressive electric locomotives we see today. In all cases they have been devised so that man can perform his tasks with the minimum effort.

In all machines there are two main parts, the **input side** and the **output side.** We shall use the term 'load' when we are speaking about the force which the machine has to overcome at the output side. It is in fact a resistance in most cases, but since there must always be some form of resistance within the machine, it will be less confusing if we speak of the external resistance which the machine is intended to overcome as the 'load'.

It follows from this definition that since the load is a force, its value will be expressed in newtons. This is the same unit which will be used to measure the effort required at the input side to overcome the load.

The relationship between the load and the effort depends upon the type of machine being used. It gives a good indication of the advantage to be gained by using the machine. The relationship is called the **mechanical advantage** of the machine.

$$\text{Mechanical advantage} = \frac{\text{load}}{\text{effort}}.$$

If you have seen a casting being lifted into position by means of a pulley block tackle you will have noticed that the casting moves very slowly compared with the rate at which the operator moves the chain on the input side. If you watch the rotation of the shafts of a worm gear box you will see that the output shaft, or the wheel shaft, rotates much more slowly than the input or worm shaft. The ratio between the distances or the angles moved through by the effort at the input side and the load at the output side is the **velocity ratio** of the machine.

$$\text{Velocity ratio} = \frac{\text{distance moved by effort}}{\text{distance moved by load in same time}}.$$

When we consider the amount of work which is 'put into' the machine by the effort on the input side, and also the amount of work which is 'got out' of the machine by the load on the output side we shall have to introduce another expression—the **efficiency** of the machine. As the effort moves through any given distance it does a definite amount of work. During this same interval of time, the load is being moved through a certain distance. This is only possible if the machine does the necessary amount of work on the load. The efficiency is the ratio of these two amounts of work. Were the machine perfect (which means there were no friction at the bearings, no drag due to windage or the viscous effect of lubricants and, if, in the case of a lifting block, the hook and the chain supporting the hook had no weight so that you had only the load to lift and not the parts of the pulley block in addition), then you could expect to get the same amount of work out of the machine as you had put into it.

A diagrammatic representation of the essential parts of a machine is shown in Fig. 11.1. Lever A on the input side is connected to lever B on the output side by means of a system of levers, pulleys, and gears inside the machine. We are not for the moment concerned with the connections between the input and the output sides because these do not affect the basic calculations in connection with the machine. When a load W N is hung on the end of lever B, it is found that an effort of P N is required at the end of lever A in order to lift the load. This means that under these conditions the mechanical advantage of the machine is W/P. That is you can lift a load of W/P times the effort which you are applying. You will notice however that the load moves much more slowly than the effort. Let the input lever be moved so that the point at which the effort is applied moves through a distance x. It will be seen that the point at which the load is applied on the output lever has moved through a distance h. This means that the velocity ratio of the machine is equal to x/h, i.e. the effort moves x/h times as far as the load moves in the same time.

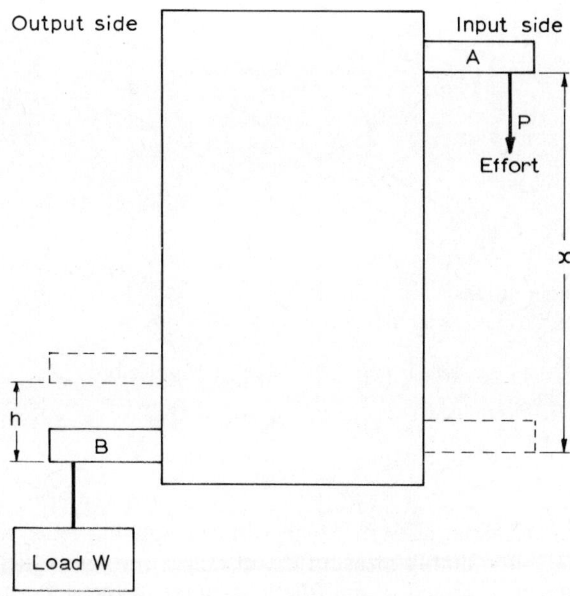

Figure 11.1 *The Essentials of a Machine*

Basic formulae for machines

$$\text{Velocity ratio} = \frac{\text{distance moved by effort}}{\text{distance moved by load}}$$

$$\text{VR} = \frac{x}{h} \text{ (constant)}$$

$$\text{Mechanical advantage} = \frac{\text{load}}{\text{effort}}$$

$$\text{MA} = \frac{W}{P} \text{ (variable)}$$

$$\text{Mechanical efficiency} = \frac{\text{mechanical advantage}}{\text{velocity ratio}} \text{ (variable)}$$

$$\text{Mechanical efficiency} = \frac{\text{ideal effort}}{\text{actual effort}} \text{ (variable)}$$

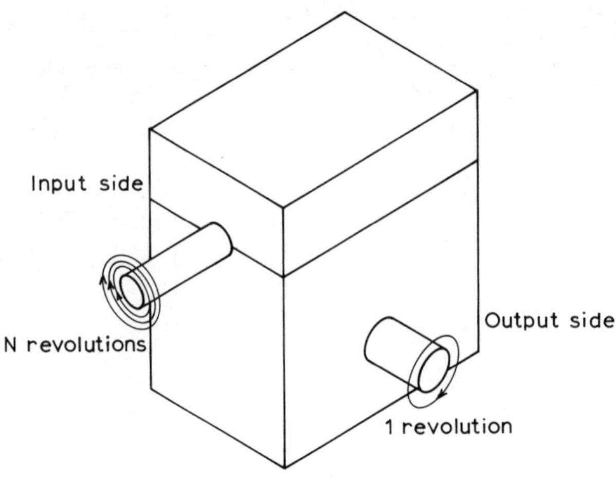

Figure 11.2 *Velocity ratio* = N

This is one of the ways in which you can determine the velocity ratio of any machine. Simply measure the distances through which the points of application of the effort and the load move in the same time and calculate the ratio x/h.

If the machine has two shafts in place of the levers, then notice how many revolutions the input shaft makes while the output shaft makes one revolution. The velocity ratio will then be:

$$\frac{\text{revolutions of input shaft}}{\text{revolutions of output shaft in the same time.}}$$

In both cases the velocity ratio can be obtained without any load being applied. *The velocity ratio depends only on the arrangement of the machine.* After the machine has been made the velocity ratio could be determined and its value stamped on the nameplate. It would remain constant provided there was no change in the arrangement of the machine. This is not the case with the mechanical advantage, as we shall see later.

Now consider the work which is done by the effort and on the load in the machine shown in Fig. 11.2.

$$\text{Work done by effort} = \text{force} \times \text{distance} = Px$$
$$\text{Work done in lifting load} = \text{force} \times \text{distance} = Wh$$

$$\text{Efficiency of machine} = \frac{\text{work done in moving load}}{\text{work done by the effort}}$$

$$= \frac{Wh}{Px} = \frac{W/P}{x/h}$$

$$\therefore \quad \text{Efficiency} = \frac{\text{mechanical advantage}}{\text{velocity ratio}}.$$

Efficiency is usually expressed as a percentage and the value obtained in the above calculation should therefore be multiplied by 100 to give a percentage.

If the machine were perfect, and no work had to be done in overcoming friction and moving additional masses in the form of levers and hooks, then we could expect that the work done by the effort would be equal to the work done in moving the load. This would give an efficiency of 1 or 100 per cent. and this in turn would mean that:

$$\frac{\text{mechanical advantage}}{\text{velocity ratio}} = 1 \text{ for the perfect machine.}$$

In other words, *in the perfect machine the mechanical advantage is equal to the velocity ratio.*

Consider the machine in Fig. 11.2 to be perfect. This would mean that we should require a smaller effort to move the same load, because there is no friction to overcome. Let us call this effort the ideal effort and represent it by I.

$$\text{Mechanical advantage for the perfect machine} = \frac{W}{I}$$

$$\text{Velocity ratio} = \text{mechanical advantage for perfect machine}$$

$$\therefore \quad \text{velocity ratio} = \frac{W}{I}.$$

$$\text{Ideal effort } I = \frac{\text{load}}{\text{velocity ratio}}.$$

Substituting this expression into that which we have obtained for the general machine, we have:

$$\text{Efficiency} = \frac{\text{mechanical advantage}}{\text{velocity ratio}} = \frac{W/P}{W/I} = \frac{I}{P}.$$

$$\text{Efficiency} = \frac{\text{ideal effort}}{\text{actual effort}}.$$

The law of a machine

If we carry out an experiment on a machine to find the effort required to overcome various loads, the results that we should obtain would be similar to the following:

Velocity ratio of machine = 2

load W N =	20	40	50	72	96
effort P N =	15	24	28	41	51

These results, plotted in Fig. 11.3 show that the relationship between the effort and the load can be represented by a straight line. We can express this relationship in the form of a mathematical equation:

$$P = aW + b$$

where P is the effort in newtons,
W is the load in newtons,
a and b are constants which can be obtained from the graph:
a is the slope of the graph,
b is the value of P where the line cuts the axis of P.

This equation is called the law of the machine. After it has been determined experimentally, it can be used to estimate the effort which would be required to lift any load on the machine.

In the case we are considering, the line cuts the P axis at the point 4. This is the value of b. The slope of the line is given by $\frac{mn}{pn}$ where m and p are any points on the line, mn and pn being vertical and horizontal lines.

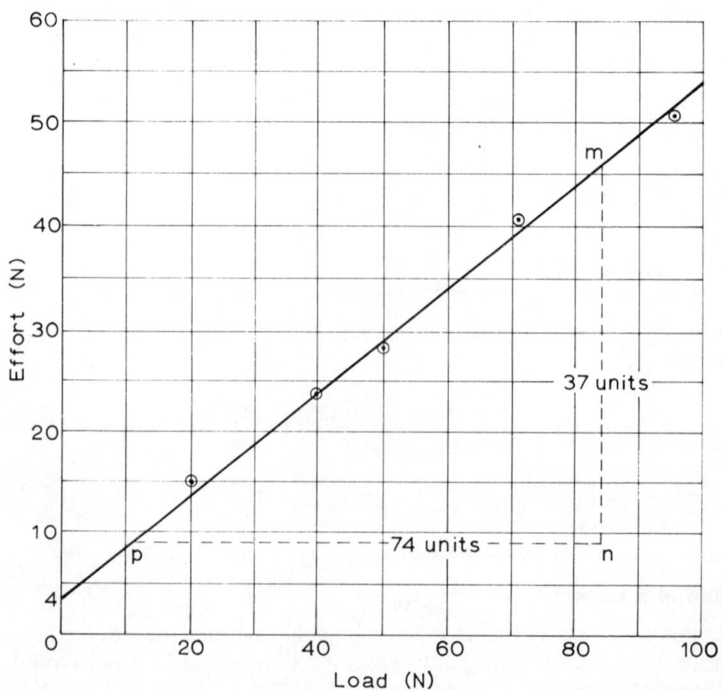

Figure 11.3 *Load-effort Graph*

From the graph $mn = 37$
$pn = 74$

$$\text{slope of line} = \frac{37}{74} = 0{\cdot}5. \quad \text{This is the value of } a.$$

The law of the machine is

$$P = 0{\cdot}5W + 4.$$

To move a load of 200 N, the required effort is given by

$$P = (0{\cdot}5 \times 200) + 4$$
$$= 104 \text{ N}.$$

Variation of Efficiency

The velocity ratio of a machine is constant and cannot be altered without making a change to the arrangement of the machine. The mechanical advantage of the machine varies with the applied load. In the results just quoted, we find that the mechanical advantage was 1·33 when the load was

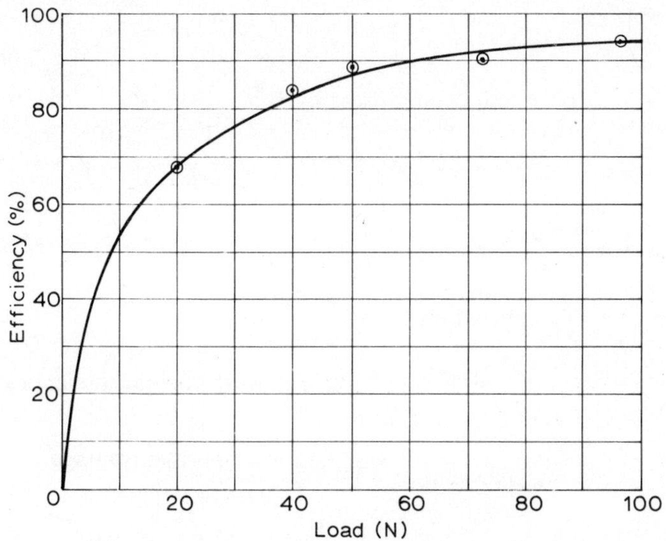

Figure 11.4 *Load-efficiency Graph*
The efficiency of a machine is not constant but increases as the load increases.

20 N. When the load was 96 N, the mechanical advantage increased to 1·88. Since efficiency is found by dividing mechanical advantage by velocity ratio, it follows that efficiency will depend upon load. The efficiency of a

machine is not constant; it will generally be found to increase with increase of load.

The mechanical advantage and the efficiency of the machine used in the above test are tabulated below and plotted in Fig. 11.4.

load N	20	40	50	72	96
effort N	15	24	28	41	51
mechanical advantage	1·33	1·66	1·76	1·78	1·88
mechanical efficiency	0·67	0·83	0·88	0·89	0·94

With the smaller loads there is a considerable increase in efficiency for a corresponding increase in load. Gradually the curve flattens out, showing that further increases in load produce only small changes in the efficiency. In fact the efficiency reaches a limiting value.

Limiting Efficiency

$$\text{Law of machine } P = aW + b$$

$$\text{Mechanical advantage} = \frac{W}{P}$$

$$= \frac{W}{aW + b}.$$

Dividing numerator and denominator by W:

$$\text{Mechanical advantage} = \frac{1}{a + (b/W)}.$$

As W increases, $\dfrac{b}{W}$ decreases, so that the value of the mechanical advantage gradually approaches the value $\dfrac{1}{a}$:

If v is the velocity ratio, then the efficiency of the machine approaches a limiting value given by:

$$\text{Limiting efficiency} = \frac{\text{limiting mechanical advantage}}{\text{velocity ratio}}$$

$$= \frac{1/a}{v}$$

$$= \frac{1}{av}.$$

This efficiency will only be reached when the load is infinite. In any case, the efficiency of the machine cannot exceed the value given above.

Example 11.1

In a certain lifting machine, an effort of 20 N is required to lift a load of 180 N. It is found that the effort has to move a distance of 900 mm while the load moves 60 mm. Find:

(*a*) the work done in lifting the load 1 m,
(*b*) the distance moved by effort while load moves 1 m,
(*c*) the work done by the effort while the load is moving 1 m,
(*d*) the efficiency of the machine,
(*e*) the effort required to lift the 180 N load if the machine were perfect.

(*a*) Work done in lifting 180 N through 1 m = force × distance
= 180 × 1
= 180 Nm.

(*b*) $$\text{Velocity ratio} = \frac{\text{distance moved by effort}}{\text{distance moved by load in same time}} = \frac{900 \text{ mm}}{60 \text{ mm}} = 15$$

i.e. effort moves 15 times distance moved by load in same time.
Distance moved by effort while load moves 1 m = 15 × 1
= 15 m

(*c*) Work done by effort = force × distance
= 20 × 15
= 300 N m.

(*d*) $$\text{Efficiency of machine} = \frac{\text{work done in lifting load}}{\text{work done by effort in same time}}$$
$$= \frac{180}{300}$$
$$= 0{\cdot}6 \text{ or } 60 \text{ per cent.}$$

(*e*) $$\text{Efficiency} = \frac{\text{ideal effort}}{\text{actual effort}}$$

$$\therefore\ 0{\cdot}6 = \frac{I}{20} \qquad \therefore\ \text{ideal effort} = 20 \times 0{\cdot}6 = 12 \text{ N.}$$

NOTE. 1. The ideal effort is 12 N and the actual effort is 20 N. This means that 8 N is required to overcome friction forces and also to lift the hook and pulley block. This is the reason for the low efficiency of 60 per cent.
2. The mechanical advantage of the machine when raising 180 N is equal to $\frac{180}{20} = 9$. We could have used this figure to find the efficiency because

$$\text{efficiency} = \frac{\text{mechanical advantage}}{\text{velocity ratio}}$$
$$= \frac{9}{15} = 0{\cdot}6.$$

This provides a useful check on the method used above.

Example 11.2

A winch is used to draw a cable through an underground conduit. The winch is designed so that the effort moves through a distance of 4 m whilst the cable is moved 100 mm.

(i) What effort would be required to produce a force of 2 kN on the cable if the winch were frictionless?
(ii) If the actual effort required is 80 N what is the efficiency of the machine?
(iii) If by oiling the winch the efficiency is increased by 12·5 per cent., what effort would then be required to produce the same force on the cable?

$$\text{Velocity ratio} = \frac{\text{distance moved by effort}}{\text{distance moved by load in same time}}$$

$$= \frac{4 \times 1000}{100} \text{ (convert both to metres)}$$

$$= 40.$$

$$\text{(i) Ideal effort} = \frac{\text{load}}{\text{velocity ratio}} \text{ (for perfect machine)}$$

$$= \frac{2 \times 1000}{40}$$

$$= 50 \text{ N}.$$

$$\text{(ii) Efficiency} = \frac{\text{ideal effort}}{\text{actual effort}}$$

$$= \frac{50}{80} = 0{\cdot}625 \text{ or } 62{\cdot}5 \text{ per cent.}$$

$$\text{(iii) New efficiency} = 62{\cdot}5 + 12{\cdot}5 = 75 \text{ per cent. or } 0{\cdot}75$$

$$\text{Actual effort} = \frac{\text{ideal effort}}{\text{efficiency}}$$

$$= \frac{50}{0{\cdot}75} = 66{\cdot}7 \text{ N}.$$

Example 11.3

Fig. 11.5 is a diagram of an experimental crab winch from which the following results were obtained:

Load in N	300	400	600	800
Effort in N	62·5	71·5	92·5	117·5

Plot a graph to show how the efficiency of the winch varies with the load moved.

[N.C.T.E.C.]

Let the handle spindle make one revolution. Then the effort which is applied at a radius of 400 mm moves through a distance of radius $\times\ 2\pi$, which equals 800π mm.

The drum shaft makes $1 \times \frac{16}{48} = \frac{1}{3}$ rev.

The load is moved through a distance equal to $\frac{1}{3} \times 300\pi$, which equals 100π mm.

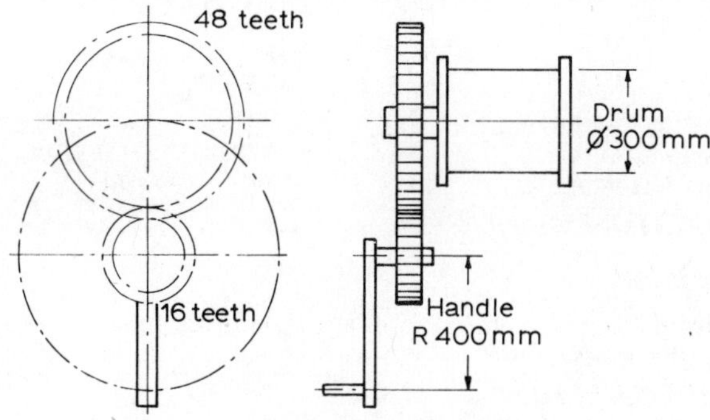

Figure 11.5

$$\text{Velocity ratio of winch} = \frac{\text{distance moved by effort}}{\text{distance moved by load in same time}}$$

$$= \frac{800\pi}{100\pi}$$

$$= 8.$$

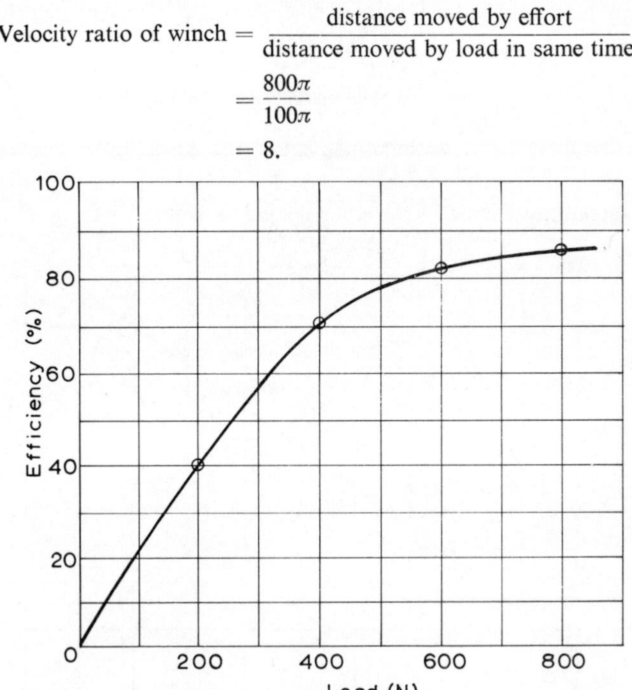

Figure 11.6

The following table can now be constructed:

$$\text{Mechanical advantage} = \frac{\text{load}}{\text{effort}}$$

$$\text{Efficiency} = \frac{\text{mechanical advantage}}{\text{velocity ratio}} = \frac{\text{MA}}{8}$$

load, N	200	400	600	800
effort, N	62·5	71·5	92·5	117·5
mechanical advantage	3·2	5·64	6·5	6·8
efficiency (fraction)	0·4	0·705	0·81	0·85
efficiency (percentage)	40	70·5	81	85

The graph is drawn in Fig. 11.6.

Example 11.4

The law of a machine as taken from a load-effort graph is $W = 3P - 20$ where W and P are in newtons.

Draw the graph and find:

(*a*) the effort required to lift a load of 70 N,
(*b*) the load lifted by an effort of 50 N,
(*c*) the effort required to operate the machine at no-load.

The graph will be a straight line so that we shall require only three points in order to draw it. (The third point is taken as a check on the other two.)

$$\text{Let } P = 10 \text{ N then } W = (3 \times 10) - 20 = 10 \text{ N}$$
$$P = 60 \text{ N then } W = (3 \times 60) - 20 = 160 \text{ N}$$
$$P = 40 \text{ N then } W = (3 \times 40) - 20 = 100 \text{ N.}$$

Using these three values we can draw the graph. (Fig. 11.7.)

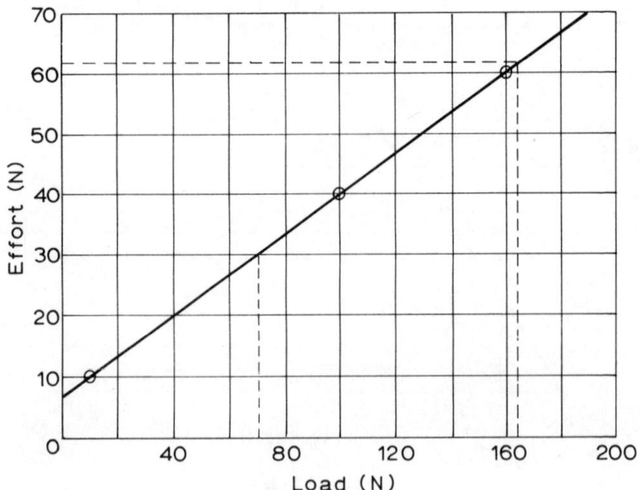

Figure 11.7

From the graph the answers to (*a*), (*b*), and (*c*) can be obtained.

Effort required to lift load of 70 N is 30 N.
Load lifted by effort of 50 N is 130 N.
Effort required to operate machine at no-load (i.e. when $W = 0$) is 6·6 N.

These values may also be obtained by substituting in the equation for the machine.

Example 11.5

In a test on a certain machine, it is found that a force of 30 N is required to lift a load of 210 N and a force of 50 N is required to lift a load of 380 N. Assuming that the graph of effort plotted against load is a straight line, determine the probable load that would be lifted by a force of 100 N.

If the velocity ratio of the machine is 20, calculate the efficiency of the machine at each load. What is the limiting efficiency of the machine?

Let the equation, connecting load and effort be $P = aW + b$

where a and b are constants,
P is the effort in N,
W is the load in N.

When W is 210, P is 30

$$\therefore \quad 30 = 210a + b \qquad (1)$$

W is 380, P is 50

$$\therefore \quad 50 = 380a + b \qquad (2)$$

The solution of these two simultaneous equations will give the value of a and b.

Subtracting (1) from (2), $20 = 170a$

$$a = \frac{1}{8{\cdot}5}.$$

Substituting into (1), $30 = \dfrac{210}{8{\cdot}5} + b$

$$30 = 24{\cdot}1 + b$$
$$b = 5{\cdot}9.$$

Law of the machine is $P = \dfrac{W}{8{\cdot}5} + 5{\cdot}9.$

Load raised by 100 N effort:

$$100 = \frac{W}{8{\cdot}5} + 5{\cdot}9$$
$$W = 8{\cdot}5 \times 94{\cdot}1$$
$$= 800 \text{ N}.$$

Efficiencies

(*a*) Effort 30 N
Load 210 N

Mechanical advantage $= \frac{210}{30} = 7$

$$\text{Efficiency} = \frac{\text{mechanical advantage}}{\text{velocity ratio}} = \frac{7}{20}$$
$$= 0{\cdot}35 \text{ or } 35 \text{ per cent.}$$

(*b*) Effort 50 N
Load 380 N

Mechanical advantage $= \frac{380}{50} = 7{\cdot}6$

$$\text{Efficiency} = \frac{\text{mechanical advantage}}{\text{velocity ratio}} = \frac{7{\cdot}6}{20}$$
$$= 0{\cdot}38 \text{ or } 38 \text{ per cent.}$$

(*c*) Effort 100 N
Load 800 N

$$\text{Mechanical advantage} = \tfrac{800}{100} = 8$$

$$\text{Efficiency} = \frac{\text{mechanical advantage}}{\text{velocity ratio}} = \frac{8}{20}$$

$$= 0{\cdot}4 \text{ or } 40 \text{ per cent.}$$

The limiting efficiency

As the load increases, the mechanical advantage approaches a limiting value given by $1/a$, where a is the constant in the equation for the machine.

$$\therefore \quad \text{Limiting mechanical advantage} = \frac{1}{1/8{\cdot}5} = 8{\cdot}5$$

$$\text{Limiting efficiency} = \frac{\text{mechanical advantage}}{\text{velocity ratio}}$$

$$= \frac{8{\cdot}5}{20}$$

$$= 42{\cdot}5 \text{ per cent.}$$

The load lifted by the 100 N effort is 800 N. The efficiency of the machine at each load is 35 per cent., 38 per cent. and 40 per cent. The limiting efficiency of the machine is 42·5 per cent. Whatever load is applied to the machine, the efficiency cannot exceed this amount.

Problems dealing with machines usually involve the determination of the velocity ratio. It is possible to obtain standard formulae for each type of machine, but in general this approach is not recommended. The most satisfactory method is to obtain the velocity ratio for the particular machine working from first principles. There are two basic methods of calculating this velocity ratio:

(1) Given the details of the machine, let the load move through unit distance (such as one turn of a wheel or 1 m linear) and calculate the distance through which the effort moves. The velocity ratio is then obtained from the formula

$$\text{Velocity ratio} = \frac{\text{distance moved by effort}}{\text{distance moved by load in same time}}.$$

(Incidentally, it may be easier in certain cases to let the effort move through unit distance and calculate the distance moved by the load. There are a few cases, particularly pulley-block systems, in which this second method is more easily followed than the first.)

(2) Assume the machine to be perfect, in which case the velocity ratio is equal to the mechanical advantage. Obtain a relationship between the load and the effort to calculate the mechanical advantage and then equate this to the velocity ratio.

Both methods are illustrated in the following problems.

The lever

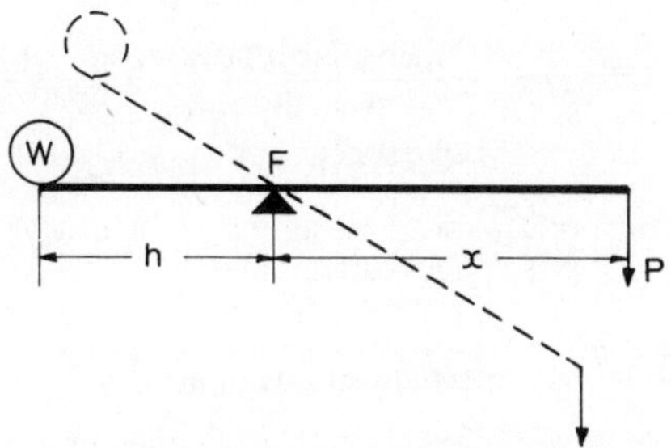

Figure 11.8 *Simple Lever*
Fulcrum between the load and the effort.

The lever, shown in Fig. 11·8, is arranged with the fulcrum F between the load W and the effort P. Assuming that the lever is a perfect machine, i.e. the lever has no mass and there are no friction forces acting at the fulcrum or the points of application of the load and effort, we can find the relationship between the load and the effort by taking moments about the fulcrum.

$$Px = Wh.$$

$$\text{Mechanical advantage} = \frac{\text{load}}{\text{effort}} = \frac{W}{P}.$$

In a perfect machine, the velocity ratio is equal to the mechanical advantage.

$$\text{Velocity ratio} = \frac{W}{P} = \frac{x}{h}.$$

In practice, for a given load, P will be greater than the value given by the equation $Px = Wh$. The velocity ratio, however, will always be $\frac{x}{h}$.

Example 11.6

A lever of the type shown in Fig. 11·8 is 600 mm long and is mounted on a pivot 200 mm from one end. An effort of 8 N is applied at the end of the longer arm to overcome a load of 15 N acting at the other end. What is the velocity ratio, mechanical advantage, and the efficiency of the lever? If the lever were mounted on a frictionless pivot, what effort would be required to lift the load?

$$\text{Velocity ratio } \frac{x}{h} = \frac{600 - 200}{200} = 2$$

$$\text{Mechanical advantage} = \frac{\text{load}}{\text{effort}} = \frac{15}{8} = 1{\cdot}875$$

$$\text{Efficiency} = \frac{\text{mechanical advantage}}{\text{velocity ratio}} = \frac{1{\cdot}875}{2}$$

$$= 0{\cdot}9375 \text{ or } 93\tfrac{3}{4} \text{ per cent.}$$

In a perfect machine, mechanical advantage = velocity ratio.

$\therefore$ Mechanical advantage = 2.

$$\text{Ideal effort} = \frac{\text{load}}{\text{mechanical advantage}} = \frac{15}{2} = 7\tfrac{1}{2} \text{ N.}$$

Note that this would be obtained by taking moments about F.

$$P \times 400 = 15 \times 200$$

$$P = \frac{15 \times 200}{400} = 7\tfrac{1}{2} \text{ N.}$$

Velocity ratio is 2, mechanical advantage is 1·875, efficiency is 93·75%, and the ideal effort to lift 15 N is 7½ N.

Alternative lever arrangements

In the type shown in Fig. 11.9, the load W is applied between the effort P and the fulcrum F.

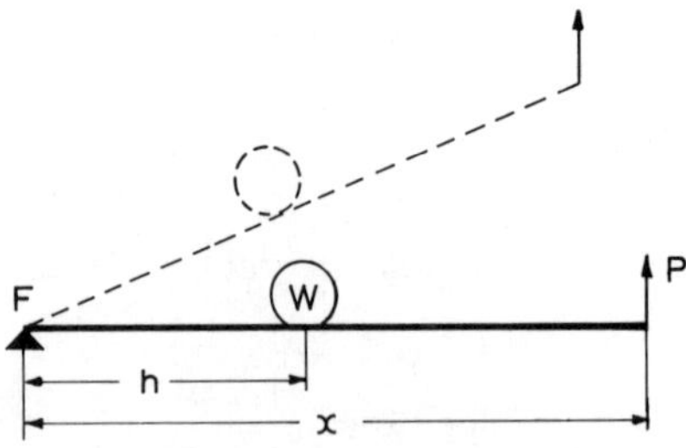

Figure 11.9 *Simple Lever*
Load between the fulcrum and the effort.

The velocity ratio is again x/h where both distances are measured from the fulcrum.

Bell crank levers

A lever which is bent as shown in Fig. 11.10 and Fig. 11.11 is called a bell crank lever. It is used to provide a change in the direction of the applied forces.

Again the velocity ratio is x/h.

Note that in all the cases considered, x and h are the *perpendicular* distances from the fulcrum to the line of action of the effort and load.

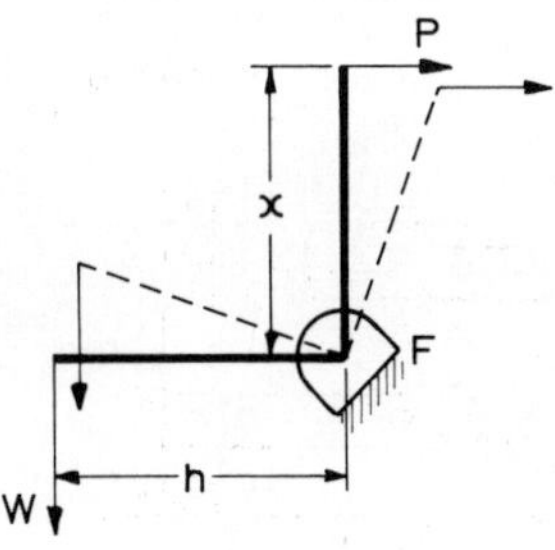

Figure 11.10 *Bell Crank Lever*

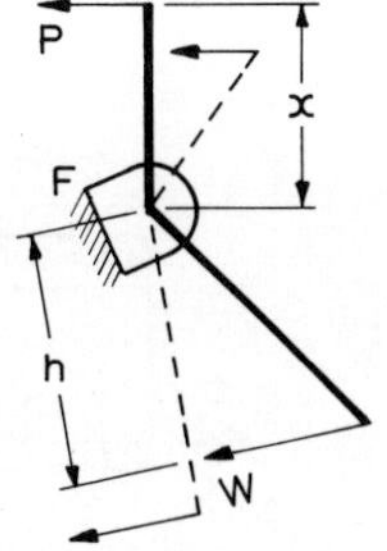

Figure 11.11 *Bell Crank Lever*

Combination of levers

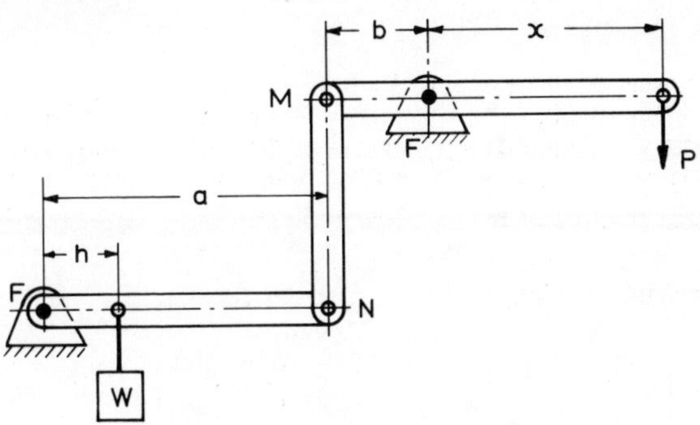

Figure 11.12 *Combination of Levers*

Two lever arrangements are connected together by the link MN (Fig. 11.12). The overall velocity ratio of the system can be found by considering W to move a distance d. Then link MN will move through a distance given by:

$$d \times \frac{a}{h}$$

and P will move a distance of $d \times \frac{a}{h} \times \frac{x}{b}$.

Overall velocity ratio is therefore $\frac{a}{h} \times \frac{x}{b}$

which is equal to the product of the velocity ratios of the separate levers.

Example 11.7

In the machine shown diagrammatically in Fig. 11.13 a load of 4800 N at A is overcome by an effort of 3 N at M. What is the efficiency of the machine?

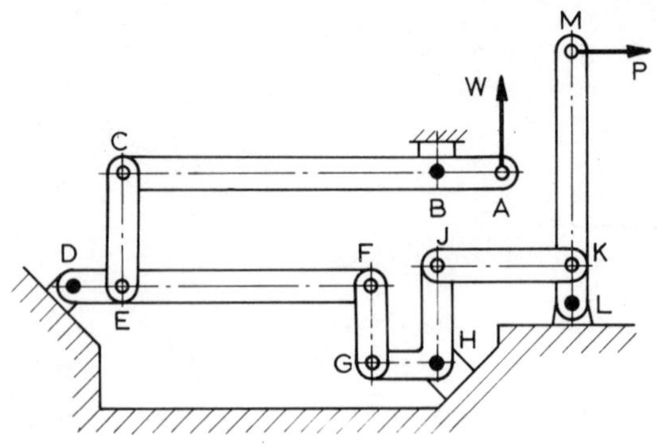

Figure 11.13

The dimensions of the levers are as follows:

ABC pivoted at B	AB = 20 mm	BC = 400 mm
DEF pivoted at D	DE = 30 mm	EF = 270 mm
GHJ pivoted at H	GH = 80 mm	HJ = 160 mm
LKM pivoted at L	LK = 30 mm	LM = 180 mm

$$\text{Lever ABC velocity ratio} = \frac{400}{20} = 20$$

$$\text{DEF} \quad ,, \quad ,, \quad = \frac{270 + 30}{30} = \frac{300}{30} = 10$$

$$\text{GHJ} \quad ,, \quad ,, \quad = \frac{160}{80} = 2$$

$$\text{LKM} \quad ,, \quad ,, \quad = \frac{180}{30} = 6$$

$$\text{Overall velocity ratio} = 20 \times 10 \times 2 \times 6 = 2400$$

$$\text{Mechanical advantage} = \frac{\text{load}}{\text{effort}} = \frac{4800}{3} = 1600$$

$$\text{Efficiency of machine} = \frac{\text{mechanical advantage}}{\text{velocity ratio}} = \frac{1600}{2400} = \tfrac{2}{3}$$

$$= 66\tfrac{2}{3} \text{ per cent.}$$

The efficiency of the machine is $66\frac{2}{3}$ per cent.

Wheel and axle (Fig. 11.14)

The load is raised by a cord wrapped around the axle whose diameter is *d*. The diameter *D* of the wheel is larger than that of the axle to which it is rigidly connected. The effort is applied to a cord wrapped round the wheel. The whole arrangement is mounted in bearings.

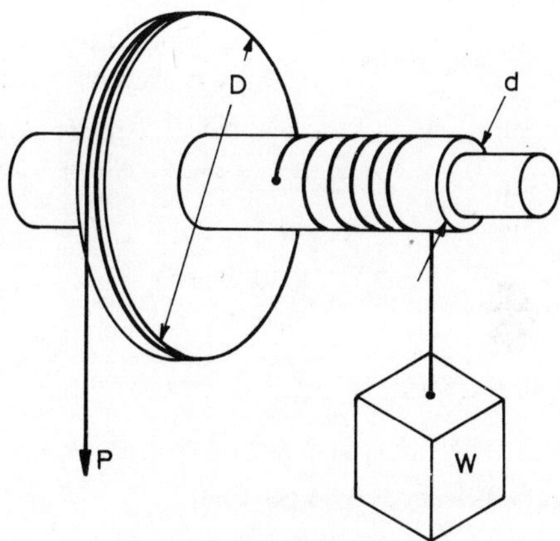

Figure 11.14 *Wheel*

When the wheel and axle make one revolution, a length of cord equal to πd is wound on to the axle, i.e. the load is raised through a distance πd. Similarly, the effort moves through a distance πD. The effort will, of course, move away from the wheel.

$$\text{Hence velocity ratio} = \frac{\text{distance moved by effort}}{\text{distance moved by load in same time}}$$

$$= \frac{\pi D}{\pi d}$$

$$= \frac{D}{d}$$

$$\text{Mechanical efficiency} = \frac{\text{work done in lifting load}}{\text{work done by effort in same time}}$$

$$= \frac{\pi d \times W}{\pi D \times P}$$

$$= \frac{Wd}{PD}.$$

Example 11.8

A wheel and axle are used for raising water from a well. The diameter of the axle is 250 mm and the crank is 600 mm long. An effort of 30 N is required to raise a load of 100 N. What is the mechanical efficiency of the arrangement?

$$\text{Velocity ratio} = \frac{\text{distance moved by effort}}{\text{distance moved by load in same time}}$$

$$= \frac{2\pi \times 600}{\pi \times 250}$$ since equivalent diameter of the wheel would be twice radius of crank $= 2 \times 600$ mm

$$= 4{\cdot}8$$

$$\text{Mechanical advantage} = \frac{\text{load}}{\text{effort}} = \frac{100}{30} = 3\tfrac{1}{3}$$

$$\text{Mechanical efficiency} = \frac{\text{MA}}{\text{VR}} = \frac{3\frac{1}{3}}{4{\cdot}8}$$

$$= 0{\cdot}694 \text{ or } 69{\cdot}4 \text{ per cent.}$$

The mechanical efficiency is 69·4 per cent.

The inclined plane

The inclined plane was one of the earliest machines used by man. It was found much easier to raise a load by moving it up an incline than by lifting it through a vertical height.

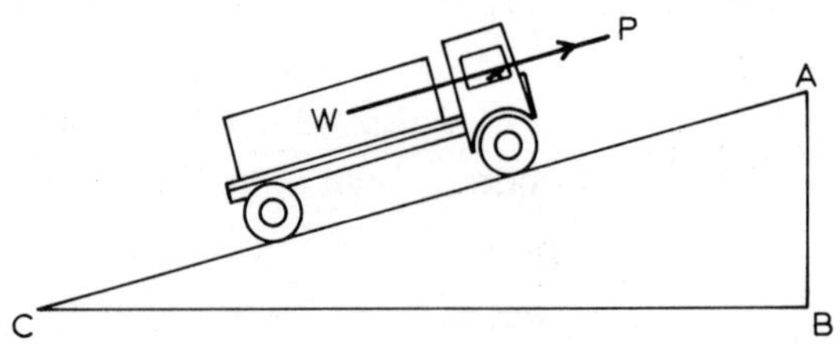

Figure 11.15 *Inclined Plane*

The lorry is in effect lifted through a vertical height AB. The tractive force moving the lorry up the plane is P acting parallel to the plane.

The effort P moves through a distance AC while the load W moves through a vertical distance equal to AB. (Notice that distances must always be measured in the direction in which the force acts. The weight of the load acts vertically downwards.)

$$\text{Velocity ratio} = \frac{\text{distance moved by effort}}{\text{distance moved by load in same time}}$$

$$= \frac{\text{AC}}{\text{AB}}$$

$$\text{Mechanical advantage} = \frac{\text{load}}{\text{effort}} = \frac{W}{P}$$

$$\text{Mechanical efficiency} = \frac{\text{work done in lifting load}}{\text{work done by effort in same time}}$$

$$= \frac{W \times \text{AB}}{P \times \text{AC}}.$$

Example 11.9

Find the mechanical advantage and velocity ratio of a perfectly smooth inclined plane when a force of 28 N parallel to the plane is sufficient to push a weight of 224 N up the plane with uniform velocity. If the surface of the plane became roughened so that the applied force had to be increased to 40 N, what would then be the mechanical advantage, the velocity ratio and the efficiency of the plane? [U.L.C.I.]

(*a*) $$\text{The mechanical advantage} = \frac{\text{load}}{\text{effort}} = \frac{224}{28} = 8.$$

Since the plane is perfectly smooth, it can be considered to be a perfect machine whose efficiency is 100 per cent. In such a case the velocity ratio is equal to the mechanical advantage.

$$\therefore \quad \text{velocity ratio} = 8.$$

(*b*) When surface is rough.

$$\text{Mechanical advantage} = \frac{\text{load}}{\text{effort}} = \frac{224}{40} = 5{\cdot}6.$$

The velocity ratio will still be 8, since it is unaffected by the roughness of the plane. It only depends upon the angle of the plane.

$$\text{Mechanical efficiency} = \frac{\text{mechanical advantage}}{\text{velocity ratio}}$$

$$= \frac{5{\cdot}6}{8}$$

$$= 0{\cdot}7 \text{ or } 70 \text{ per cent.}$$

Case (*a*) Mechanical advantage = 8
Velocity ratio = 8
Mechanical efficiency = 100 per cent.
(*b*) Mechanical advantage = 5·6
Velocity ratio = 8
Mechanical efficiency = 70 per cent.

Screw jack

Fig. 11.16 shows diagrammatically the arrangement of a simple screw jack. The three essential parts are:

(1) The screw, usually single start, provided with a collar at its upper end in which are drilled 'tommy holes.' There is also a cap at the end of the screw which is arranged so that it does not rotate with the screw. The load is taken by the cap.
(2) The main body of the jack into which the screw is screwed.
(3) The lever fitted into one of the tommy holes in the upper part of the screw.

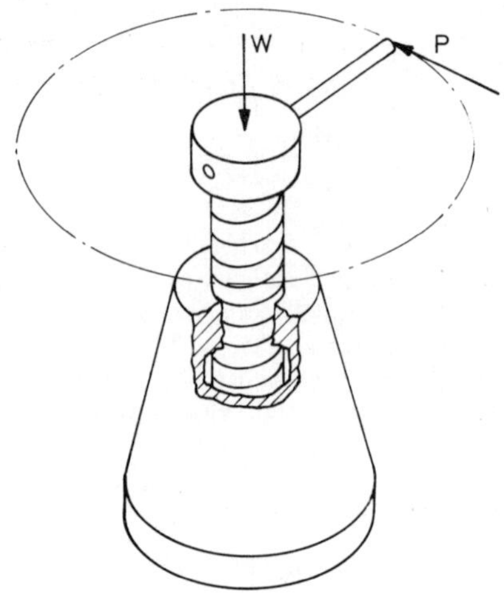

Figure 11.16 *Screw Jack*

The effort is applied at the end of the arm, length r. For each revolution of the arm the load is raised by an amount equal to the lead of the screw.

Generally the screw is used to lift loads so that the screw axis is vertical and the lever moves in a horizontal plane. The effort is applied at the end of the lever in a direction perpendicular to its length.

When the screw makes one revolution, the effort moves through a distance of $2\pi r$ where r is the length of the lever. The load moves through a vertical distance equal to the lead of the screw l, which in the case of a single start thread is equal to the pitch of the screw.

$$\text{Velocity ratio} = \frac{\text{distance moved by effort}}{\text{distance moved by load in same time}}$$

$$= \frac{2\pi r}{l}$$

$$\text{Mechanical advantage} = \frac{\text{load}}{\text{effort}}$$

$$= \frac{W}{P}$$

$$\text{Mechanical efficiency} = \frac{\text{work done in lifting load}}{\text{work done by effort in same time}}$$

$$= \frac{Wl}{2\pi r P}.$$

Example 11.10

A screw jack is operated by a bar pushed through a hole in the spindle. The pitch of the screw is 15 mm. A force of 300 N applied at right angles to the bar, at a radius of 350 mm from the axis of the screw, is found to lift a casting weighing 20 kN which rests upon the top of the jack. Find the velocity ratio, the mechanical advantage, and the mechanical efficiency of the machine under these conditions. [U.E.I.]

When the screw makes 1 revolution:
the effort moves $2\pi \times 350$ mm
the load moves 15 mm

$$\text{Velocity ratio} = \frac{\text{distance moved by effort}}{\text{distance moved by load in same time}}$$

$$= \frac{2\pi \times 350}{15} \qquad \left(\pi = \frac{22}{7}\right)$$

$$= 147.$$

$$\text{Mechanical advantage} = \frac{\text{load}}{\text{effort}}$$

$$= \frac{20 \times 1000}{300}$$

$$= 66\tfrac{2}{3}.$$

$$\text{Mechanical efficiency} = \frac{\text{work done in moving load}}{\text{work done by effort in same time}}$$

$$= \frac{20 \times 1000 \times 15}{300 \times 2\pi \times 350}$$

$$= 0{\cdot}455 \text{ or } 45{\cdot}5 \text{ per cent.}$$

Alternatively:

$$\text{Mechanical efficiency} = \frac{\text{mechanical advantage}}{\text{velocity ratio}}$$

$$= \frac{66\tfrac{2}{3}}{147}$$

$$= 0{\cdot}455 \text{ or } 45{\cdot}5 \text{ per cent.}$$

The velocity ratio is 147, the mechanical advantage is 66⅔, and the efficiency under the given conditions is 45·5 per cent.

Gear drives

In a gear wheel, a number of specially shaped projections or 'teeth' are produced on the periphery of the wheel. These teeth mesh with corresponding teeth on a second wheel so that motion can be transferred from one to the other without any slip taking place.

Two wheels A and B are geared together so that wheel A is driving wheel B in the opposite direction to A. Let the number of teeth in wheel A be a and in wheel B be b. The point on the line joining the centres of the wheels at which the gears mesh is called the pitch point P. Now in any interval of time the same number of teeth will pass point P whether we

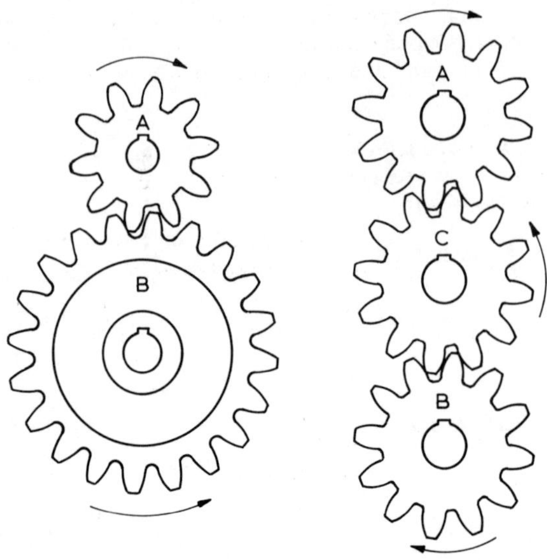

Figure 11.17 *Spur Gear Drive*

The idler wheel C has the effect of making A and B rotate in the same direction without altering the velocity ratio of the drive.

think of wheel A or B. Suppose in a given interval of time, t teeth pass point P. This means that A has made $\frac{t}{a}$ revolutions and B has made $\frac{t}{b}$ revolutions in the same time.

The velocity ratio of the drive is:

$$\frac{\text{Revolutions of driver wheel (the effort)}}{\text{Revolutions of driven wheel (the load) in same time}}$$

$$= \frac{t/a}{t/b}$$

$$= \frac{b}{a}$$

$$= \frac{\text{number of teeth in driven wheel}}{\text{number of teeth in driver wheel}}.$$

If we wish to know the speed of the driven wheel, we can apply the relationship:

$$\frac{\text{Rev/min driver}}{\text{Rev/min driven}} = \frac{\text{teeth in driven}}{\text{teeth in driver}}$$

$$\text{Rev/min driven wheel} = \frac{\text{teeth in driver}}{\text{teeth in driven}} \times \text{rev/min driver.}$$

Use of the idler wheel (Fig. 11.17)

A third wheel C, known as an idler wheel, is sometimes inserted between A and B. The effect of this is to make wheel B rotate in the same direction as that of wheel A, although the speed of B is not affected by the insertion of C.

Let the teeth in the wheels A, B and C be denoted by a, b and c respectively.

$$\text{Rev/min of C} = \frac{a}{c} \times \text{rev/min of A}$$

$$\text{Rev/min of B} = \frac{c}{b} \times \text{rev/min of C}$$

$$= \frac{c}{b} \times \frac{a}{c} \times \text{rev/min of A}$$

$$= \frac{a}{b} \times \text{rev/min of A.}$$

Example 11.11

Fig. 11.18 shows the gear drive in a certain machine tool. If shaft A rotates at 900 rev/min what is the speed of shaft K?

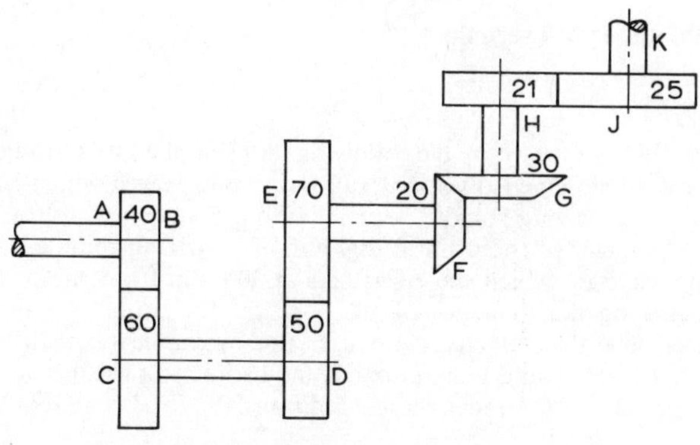

Figure 11.18

$$\text{Speed of C} = \text{speed of B} \times \frac{B}{C}$$
$$= 900 \times \frac{40}{60}$$
$$= 600 \text{ rev/min, which is also the speed of D.}$$

$$\text{Speed of E} = \text{speed of D} \times \frac{D}{E}$$
$$= 600 \times \frac{50}{70}$$
$$= 428\tfrac{4}{7} \text{ rev/min, which is also the speed of F.}$$

$$\text{Speed of G} = \text{speed of F} \times \frac{F}{G}$$
$$= 428\tfrac{4}{7} \times \frac{20}{30}$$
$$= 285\tfrac{5}{7} \text{ rev/min, which is also the speed of H.}$$

$$\text{Speed of K} = \text{speed of H} \times \frac{H}{J}$$
$$= 285\tfrac{5}{7} \times \frac{21}{25}$$
$$= 240 \text{ rev/min.}$$

A question of this type can be solved by a continuous calculation, taking the wheels in order and starting from A.

$$\text{Speed of K} = \text{speed of A} \times \frac{B}{C} \times \frac{D}{E} \times \frac{F}{G} \times \frac{H}{J}$$
$$= 900 \times \frac{40}{60} \times \frac{50}{70} \times \frac{20}{30} \times \frac{21}{25}$$
$$= 240 \text{ rev/min.}$$

Shaft K rotates at 240 rev/min.

Example 11.12

State clearly what are meant by the following terms applied to a lifting machine: (*a*) mechanical advantage, (*b*) velocity ratio. A single geared winch has a wheel of 20 teeth on the driving spindle, gearing with a 50-tooth wheel on the drum spindle. The diameter of the drum is 200 mm. The driving spindle is turned by means of a handle on which the effort acts at 300 mm from the centre of that spindle. Neglecting the diameter of the rope on the drum and the effect of friction, determine the velocity ratio and the mechanical advantage of the machine. What force would be required at the handle to raise a load of 1500 N suspended from the drum under these conditions? [N.C.T.E.C.]

A diagram of the arrangement is shown in Fig. 11.19.
Let the driving spindle make 1 revolution anticlockwise.

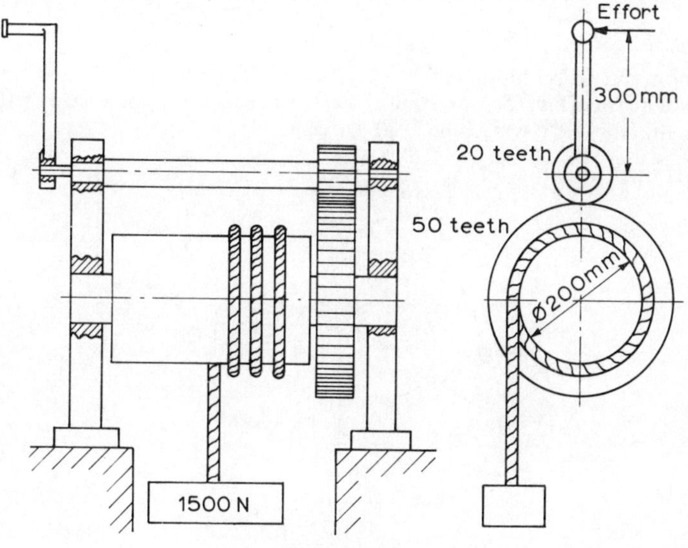

Figure 11.19

The effort at the end of the handle moves a distance equal to $2\pi \times 300 = 600\pi$ mm.

The drum make $1 \times \dfrac{20}{50} = 0{\cdot}4$ revolutions clockwise.

Thus, the load is lifted by $0{\cdot}4 \times 200\pi$ mm $= 80\pi$ mm (one turn of the drum would lift the load 200π mm, therefore 0·4 turns will lift it $0{\cdot}4 \times 200\pi$).

$$\text{Velocity ratio} = \frac{\text{distance moved by effort}}{\text{distance moved by load in same time}}$$

$$= \frac{600\pi}{80\pi}$$

$$\text{Velocity ratio} = 7\tfrac{1}{2}.$$

Since the machine is considered frictionless, the mechanical advantage equals the velocity ratio.

$$\therefore \quad \text{Mechanical advantage} = 7\tfrac{1}{2}.$$

$$\frac{\text{Load}}{\text{Effort}} = 7\tfrac{1}{2}$$

$$\therefore \quad \text{Effort} = \frac{\text{load}}{7\frac{1}{2}} = \frac{1500}{7\frac{1}{2}} = 200\ \text{N}.$$

$$\text{Velocity ratio} = 7\tfrac{1}{2}$$
$$\text{Mechanical advantage} = 7\tfrac{1}{2}$$
$$\text{Effort} = 200\ \text{N}.$$

The velocity ratio is $7\frac{1}{2}$, the mechanical advantage is $7\frac{1}{2}$, and the effort required is 200 N.

Example 11.13

A worm and wheel lifting tackle is shown in Fig. 11.20. Calculate the velocity ratio of the machine and determine the effort required to lift a load of 1000 N if the efficiency of the machine is 80 per cent.

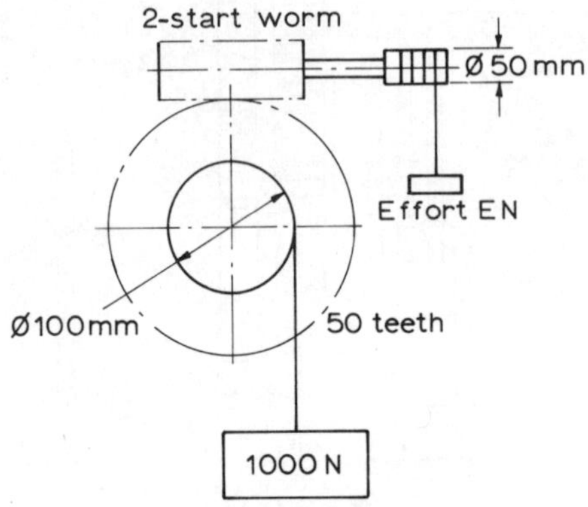

Figure 11.20 *Worm-gear Drive*

One revolution of the worm will cause the worm wheel to make $\frac{2}{50}$ rev.

For one revolution of the worm the effort will move through a distance of 50π mm.

The corresponding distance moved by the load is $\dfrac{2}{50} \times 100\pi$.

$$\text{Velocity ratio} = \frac{\text{distance moved by effort}}{\text{distance moved by load in same time}}$$

$$= \frac{50\pi}{\frac{2}{50} \times 100\pi}$$

$$= 12{\cdot}5.$$

$$\text{Mechanical advantage} = \text{velocity ratio} \times \text{efficiency}$$

$$= 12{\cdot}5 \times 0{\cdot}8$$

$$= 10$$

$$\text{Effort } E = \frac{\text{load}}{\text{mechanical advantage}}$$

$$= \frac{1000}{10}$$

$$= 100 \text{ N}$$

The velocity ratio is 12·5 and the effort to lift 1000 N is 100 N.

Belt and chain drives

Consider the two pulleys A and B, connecting which is an endless belt. A is the driving wheel. It is assumed that no slip takes place between the

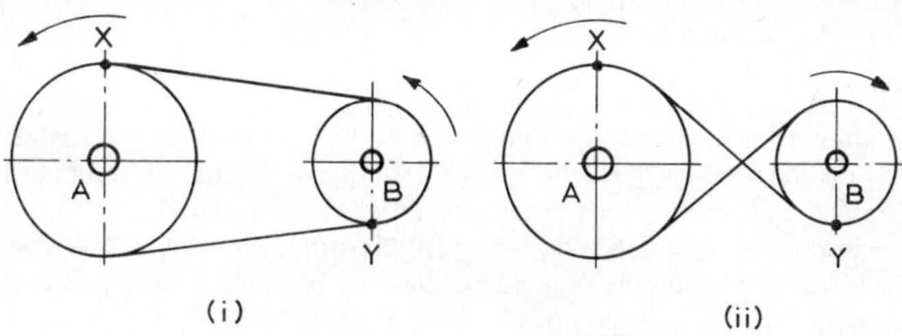

Figure 11.21 *Belt Drive*
The open belt drive shown in (i) is arranged so that A and B rotate in the same direction. If the belt is crossed as in (ii) the wheels rotate in opposite directions.

belt and the pulleys. When A makes N revolutions, the point X on the rim of A moves through a distance of $N\pi d_a$. In moving round, a length of belt equal to $N\pi d_a$ is moved from right to left and this in turn causes a point Y on the rim of wheel B to move through a distance of $N\pi d_a$.

$$\text{Number of revolutions made by wheel B} = \frac{N\pi d_a}{\text{circumference of B}}$$

$$= \frac{N\pi d_a}{\pi d_b}$$

$$= \frac{\text{diameter of A}}{\text{diameter of B}} \times N$$

$$\therefore \quad \text{speed of B} = \text{speed of A} \times \frac{\text{diameter of A}}{\text{diameter of B}}.$$

If the belt is 'open,' as shown in Fig. 11.21 (i), both wheels rotate in the same direction. If the belt is 'crossed,' as shown in Fig. 11.21 (ii), the wheels rotate in opposite directions.

The chain drive is similar to the belt drive with the exception that the motion is transmitted much more positively. The chain cannot slip around the wheels or sprockets as they are usually called. The rollers in the chain fit into specially shaped teeth cut in the sprockets.

$$\text{Speed of B} = \text{speed of A} \times \frac{\text{diameter of A}}{\text{diameter of B}}$$

and since the number of teeth in the sprockets is proportional to the diameter of the sprockets we can write:

$$\text{Speed of B} = \text{speed of A} \times \frac{\text{teeth in A}}{\text{teeth in B}}.$$

The chain drive is always 'open,' i.e. the two sprocket wheels always rotate in the same direction.

Example 11.14

The line shaft in a machine shop is driven at 150 rev/min. For a certain lathe, the pulley on the line shaft is 200 mm in diameter and that on the countershaft is 150 mm in diameter. The speed cone on the countershaft has four steps of the following diameters: 160 mm, 120 mm, 80 mm and 40 mm. The speed cone of the lathe has the following diameters: 60 mm, 100 mm, 140 mm and 180 mm. Determine the available speeds of the lathe (Fig. 11.22).

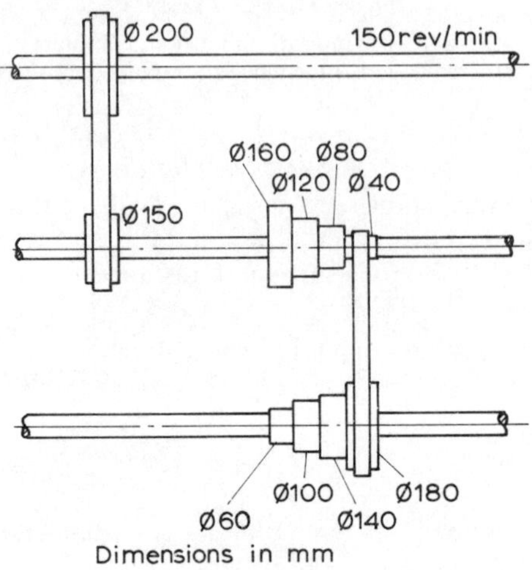

Figure 11.22

$$\text{Speed of countershaft} = \text{speed of line shaft} \times \frac{\text{driving pulley}}{\text{driven pulley}}$$

$$= 150 \times \frac{200}{150}$$

$$= 200 \text{ rev/min.}$$

Top speed of lathe. This occurs when the belt is round the 160 mm and 60 mm pulleys.

$$\text{Speed} = \text{speed of line shaft} \times \frac{\text{driving pulley}}{\text{driven pulley}}$$

$$= 200 \times \frac{160}{60}$$

$$= 533\tfrac{1}{3} \text{ rev/min.}$$

2nd speed. Belt around 120 mm and 100 mm pulleys.

$$\text{Speed} = 200 \times \frac{120}{100}$$

$$= 240 \text{ rev/min.}$$

3rd speed. Belt around 80 mm and 140 mm pulleys.

$$\text{Speed} = 200 \times \frac{80}{140}$$
$$= 114{\cdot}3 \text{ rev/min.}$$

Low speed. Belt around 40 mm and 180 mm pulleys.

$$\text{Speed} = 200 \times \frac{40}{180}$$
$$= 44{\cdot}4 \text{ rev/min.}$$

Lathe speeds are 533⅓, 240, 114·3 and 44·4 rev/min.

Example 11.15

A bicycle drive has 48 teeth on the crank sprocket and 16 on the wheel sprocket. The cranks are 200 mm long and the road wheel is 650 mm in diameter. What is the velocity ratio of the road-wheel rim to the pedal (*a*) in terms of revolutions, (*b*) in terms of linear speed?

At what speed in km/h will the bicycle be moving when the rider is pedalling at 60 turns of the crank per min?

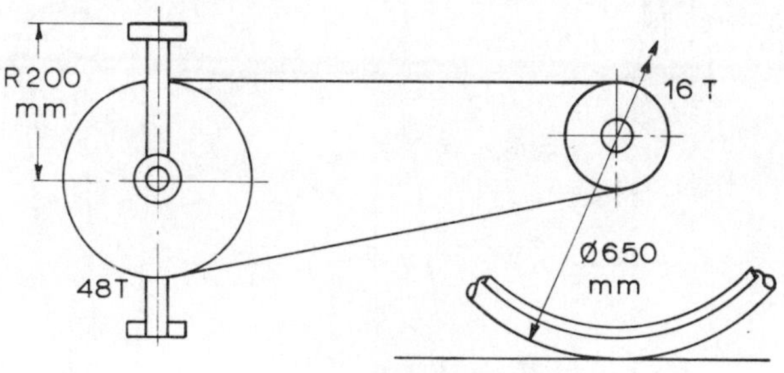

Figure 11.23

When crank makes 1 revolution the road wheel makes $\frac{48}{16} = 3$ revolutions

$\therefore$ velocity ratio (in terms of revolutions) = 3.

During one revolution, the crank travels $2\pi \times 200 = 400\pi$ mm. In the same time, the road wheel makes 3 revolutions and a point on the rim travels $3 \times \pi \times 650 = 1950\pi$ mm.

$$\therefore \quad \text{velocity ratio (in terms of linear speed)} = \frac{1950\pi}{400\pi} = 4\tfrac{7}{8}.$$

When pedalling at 60 turns per min the road wheel is making $3 \times 60 = 180$ rev/min.

$$\text{Linear speed} = 180 \times 650\pi \text{ mm/min}$$
$$= \frac{180 \times 650\pi}{10^6} \text{ km/min}$$
$$= \frac{180 \times 650\pi \times 60}{10^6} \text{ km/h}$$
$$= 22 \text{ km/h}$$

Velocity ratio is 3 in terms of revolutions and $4\frac{7}{8}$ in terms of linear speed. Linear speed is 22 km/h.

Pulley block systems

Of the machines used for lifting loads, the pulley block is perhaps the most common. The type with which we are concerned consists of a number of pulleys or sheaves freely mounted in pulley blocks. A single rope is passed over each pulley in turn. One end is fastened to either the top or the bottom pulley block, dependent upon the number of pulleys. The effort is applied to the free end of the rope and the load is attached to the lower pulley block.

The simplest arrangement of the pulley block system is shown on the left of Fig. 11.24. The single pulley serves only to reverse the direction of

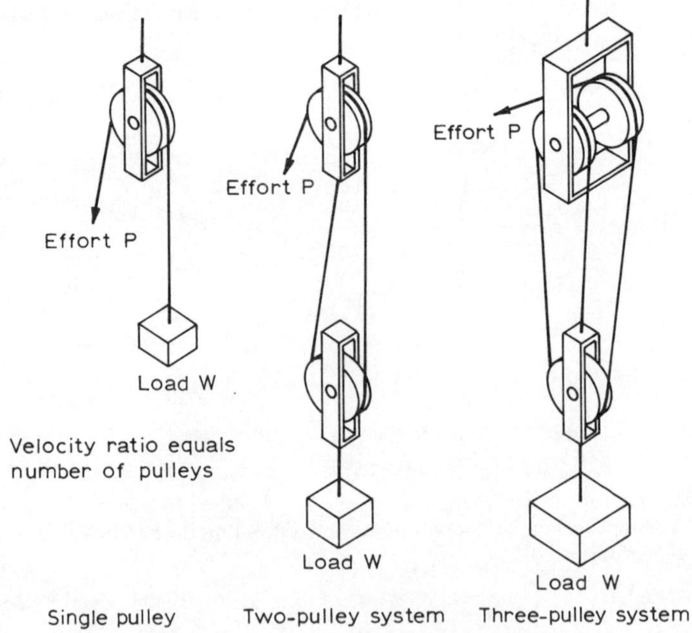

Figure 11.24 *Pulley Block Systems*

the pull. The pull in the rope is equal to the load W, and the distance travelled by the effort is equal to that through which the load moves in the same time.

In the two-pulley system, it will be seen that the load W is supported by two portions of the rope so that the tension in each portion is $W/2$. The tensions in the rope on either side of the top pulley must be equal (assuming frictionless pulley and spindle). The effort P must therefore be equal to $W/2$, which gives a mechanical advantage equal to:

$$\frac{\text{Load}}{\text{Effort}} = \frac{W}{W/2} = 2.$$

Similarly with the three-pulley system shown on the right of Fig. 11.24, the load is supported by three portions of rope, in each of which there must be a tension equal to $W/3$. Tensions on either side of the top pulley being equal gives the effort P equal to $W/3$. This means that the mechanical advantage of this arrangement is 3.

In the above considerations we have assumed perfect pulley block arrangements which involve:

(*a*) frictionless pulleys,
(*b*) weightless ropes, and
(*c*) weightless lower pulleys and blocks.

Having assumed a perfect machine, we can say that the velocity ratio is equal to the mechanical advantage.

So that: velocity ratio of 1-pulley system is 1
velocity ratio of 2-pulley system is 2
velocity ratio of 3-pulley system is 3.

In general, the velocity ratio of any pulley block system having equal diameter pulleys and a single rope is equal to the total number of pulleys.

This velocity ratio will not change when we consider the more normal case taking friction into account. This will, however, affect the mechanical advantage and the efficiency which can be calculated in the usual way.

Notice that efficiency is reduced, even though there is not the number of moving parts usually associated with machines and which would reduce efficiency, because the effort does not only have to lift the load with its pulleys, but also the lower pulley block.

Differential pulley block

Pulley block systems involving more than a total of eight pulleys are not common. It is possible to obtain a higher velocity ratio with the use of a differential pulley arrangement. Three pulleys are used. The two pulleys of slightly different diameter in the upper block are integral, i.e. they are rigidly connected together. A single pulley is mounted in the lower block connected to the load hook. A single continuous rope is used as shown in Fig. 11.25.

Again, the tension in the two portions of the rope supporting the lower block is $W/2$. These two portions of the rope act at the rim of pulleys of slightly different radii. Taking moments about the axis of the pulleys, we

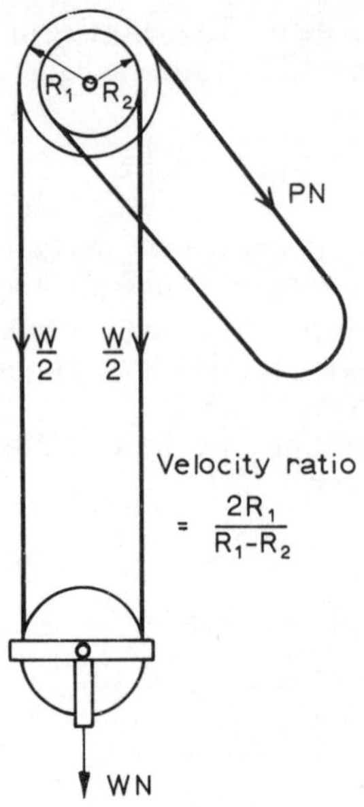

Figure 11.25 *Differential Pulley Block*

obtain the following relationship, in which R_1 and R_2 are the radii of the larger and smaller pulleys respectively:

$$PR_1 = \frac{W}{2}R_1 - \frac{W}{2}R_2$$

$$P = \frac{W}{2}\left[\frac{R_1 - R_2}{R_1}\right].$$

Since the mechanical advantage $\frac{W}{P}$ is equal to the velocity ratio in the perfect machine, we can write:

$$\text{Velocity ratio} = \frac{2R_1}{R_1 - R_2}.$$

By making $(R_1 - R_2)$ small, the velocity ratio can be made correspondingly large. This means that heavier loads can be raised by this arrangement. For this reason, a link chain is usually substituted for the rope. The arrangement is known as a **Weston pulley block** system. It is found that the chain drive is more positive than the rope drive, which relies on a frictional grip between the rope and the larger pulley. A number of 'flats' are formed round the circumference of the pulleys in the upper block into which the links of the chain fit. The pitch of the flats is the same for each pulley, and it follows that the number of flats is proportional to the circumference of the pulley which in turn is proportional to the diameter of the pulley. Hence substituting N for R in the above expression we have.

$$\text{Velocity ratio} = \frac{2N_1}{N_1 - N_2}$$

where N_1 and N_2 are the number of flats in the larger and smaller pulleys.

Example 11.16

Determine the efficiency of a six-pulley rope lifting tackle in which an effort of 100 N is required to overcome a load of 420 N.

Velocity ratio for pulley block system = number of pulleys = 6

$$\text{Mechanical advantage} = \frac{\text{load}}{\text{effort}} = \frac{420}{100}$$
$$= 4{\cdot}2$$
$$\text{Efficiency} = \frac{\text{mechanical advantage}}{\text{velocity ratio}} = \frac{4{\cdot}2}{6}$$
$$= 0{\cdot}7 \text{ or } 70 \text{ per cent.}$$

The lifting tackle has an efficiency of 70 per cent. when raising a load of 420 N.

Example 11.17

In a Weston pulley block, the larger wheel has 12 flats and the smaller wheel has 11 flats. What load can be lifted by an effort of 250 N if the efficiency at this load is 35 per cent.?

$$\text{Velocity ratio} = \frac{2N_1}{N_1 - N_2} = \frac{2 \times 12}{12 - 11}$$
$$= 24.$$
$$\text{Efficiency} = 0{\cdot}35 = \frac{\text{mechanical advantage}}{\text{velocity ratio}}$$
$$\therefore \quad \text{Mechanical advantage} = 0{\cdot}35 \times 24$$
$$= 8{\cdot}4$$
$$\therefore \quad \frac{\text{Load}}{\text{Effort}} = 8{\cdot}4$$
$$\text{Load} = 8.4 \times 250$$
$$= 2100 \text{ N.}$$

The effort of 250 N will lift a load of 2100 N.

EXERCISE 11

1 In a certain machine, an effort of 30 N is required to lift a load of 240 N. It is found that the effort moves a distance of 5 m while the load moves 0·5 m.

Find:
(*a*) the work done in lifting the load 1 m,
(*b*) the distance moved by effort while load moves 1 m,
(*c*) the work done by the effort while the load is moving 1 m,
(*d*) the efficiency of the machine,
(*e*) the effort required to lift the 240 N load if the machine were perfect.

2 A machine is designed to pull a cable through an underground conduit. The machine is arranged so that the effort moves through a distance of 6 m while the cable is pulled through 200 mm.
(*a*) What effort would be required to produce a force of 900 N on the cable if the winch were frictionless?
(*b*) If the actual effort required is 50 N, what is the efficiency of the machine?
(*c*) If by oiling the machine the efficiency is increased by 15 per cent., what effort would then be required to produce the same force on the cable?

3 A machine is used to move bars of steel in a stores. The effort moves through a distance of 8 m while the bars are pulled through 0·5 m.
(*a*) What effort would be required to produce a force of 192 N on the steel bars if the machine were perfect?
(*b*) If the actual effort required is 16 N, what is the efficiency of the machine?
(*c*) If by making a slight adjustment the efficiency of the machine is increased by 5 per cent., what effort would then be required to produce the same force of the bars?

4 In a simple screw jack, the pitch of the screw is 10 mm, and the length of the lever at the end of which the effort is applied is 360 mm. What is the velocity ratio? If an effort of 9 N applied at the end of the lever lifts a load of 1120 N, what is the efficiency?

5 A simple wheel and axle has a wheel 350 mm in diameter and an axle 100 mm in diameter. Determine the velocity ratio of the machine and the ideal and actual efforts required to lift a load of 500 N if the efficiency of the machine at this load is 80 per cent.

6 The following results were obtained from experiments with a certain type of lifting machine, whose velocity ratio was 25:

Load N	0	5	10	15
Effort N	0·094	0·45	0·81	1·17

Find the efficiency of the machine when lifting a load of 8 N.

7 Define the moment of a force.

A brake mechanism, as shown in Fig. 11.26 (i), operates with a horizontal pull of 20 N at A. The levers AP and CD turn about P and O respectively. AP = 900 mm, PB = 300 mm, OC = 300 mm. Calculate the force acting on the wheel at D. [U.E.I.]

8 In the mechanism shown in Fig. 11.26 (ii) an effort of 30 N applied at J overcomes a load of 90 N at B. The various levers are pivoted at A, E and H. AB = 100 mm, BC = 300 mm, DE = 100 mm, EG = 150 mm, HF = 75 mm, FJ = 225 mm. Determine the efficiency of the machine.

9 The efficiency of the machine shown in Fig. 11.26 (iii) is 80 per cent. The levers are pivoted about points B and E. What effort will be required at F to overcome a load of 160 N at A? AB = 500 mm, BC = 200 mm, DE = 300 mm, EF = 400 mm.

10 Fig. 11.26 (iv) shows the gearing of a machine tool. A is the motor shaft which runs at 480 rev/min. The numbers of the teeth in the wheels are as follows: B 20, C 80, D 25, E 75, F 20, G 60. G is keyed to shaft H. Determine the speed of shaft H. [U.L.C.I.]

11 An hydraulic lifting mechanism on a tipping wagon lifts a load of 20 kN through a distance of 150 mm whilst the effort moves 18 m. Calculate (*a*) the velocity ratio, (*b*) the ideal effort, and (*c*) the actual effort needed if the efficiency of the arrangement is 75 per cent.

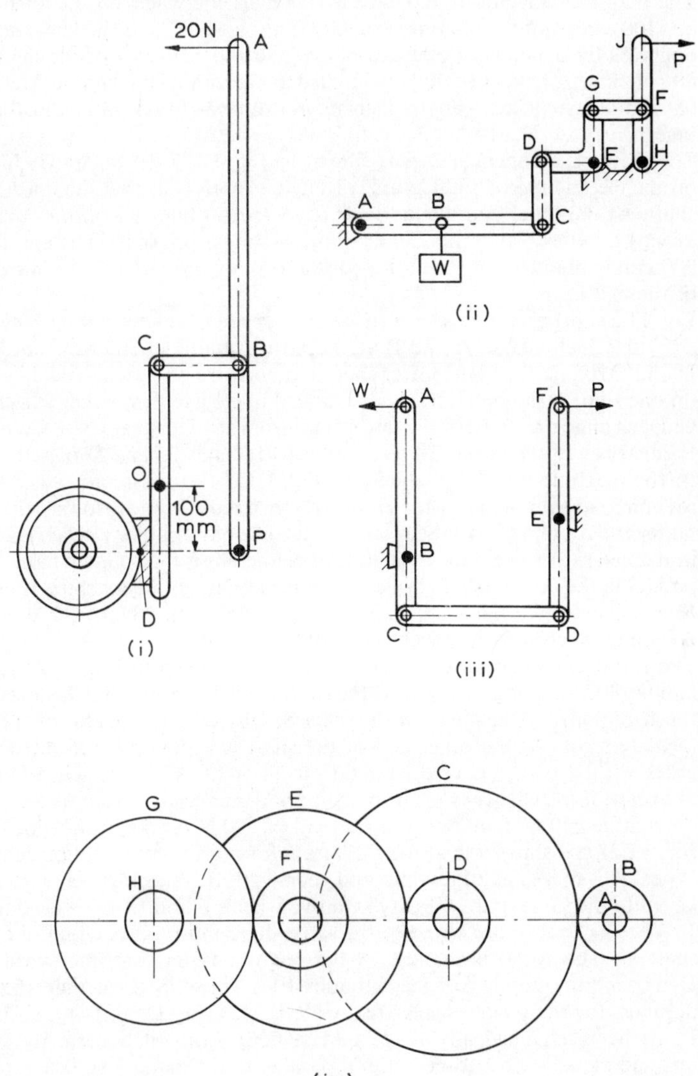

Figure 11.26

12 The pedal cranks of a bicycle are 250 mm long and are connected directly to the chain wheel of diameter 250 mm which contains 48 teeth. The sprocket on the back axle contains 16 teeth and the road wheel is 650 mm in diameter. How many rev/min must the crank be turned by the cyclist to attain a road speed of 30 km/h? If the cranks are horizontal, what is the pull in the chain when the downward force on one pedal is 40 N?

13 Fig. 11.27 (i) shows the outline diagram of a small lifting crab or winch. The barrel of the crab is 250 mm in diameter; the wheel on the barrel shaft has 100 teeth and the pinion on the driving shaft has 25 teeth. The machine is operated by a handle of 500 mm radius and the efficiency of the machine is 80 per cent. An effort of 40 N is applied at the end of the handle. Determine (*a*) the weight lifted, (*b*) the velocity ratio, and (*c*) the mechanical advantage.

14 The drive to a conveyor belt is shown in Fig. 11.27 (ii). A 20-tooth pinion on the electric motor shaft gears with a 100-tooth wheel on the intermediate shaft. At the other end of this shaft is a V-rope pulley 360 mm in diameter, driving a pulley 600 mm in diameter on the belt drum shaft. The belt drum is 300 mm in diameter. If the motor runs at 600 rev/min, what is the linear speed of the belt in m/s?

15 Fig. 11.27 (iii) gives the arrangement of a speed changing device. The upper shaft to which wheels A and B are keyed is running at 300 rev/min. Wheels D and C run freely on the lower shaft and both are provided with clutch teeth on one side. A sliding clutch E is arranged to engage with either wheels D or C. Determine the speed of the lower shaft (*a*) when E engages with C, (*b*) when E engages with D. A has 30 teeth, B has 40, C has 25 and D has 35.

16 In the machine tool drive shown in Fig. 11.27 (iv) the motor runs at 1000 rev/min but is provided with an internal speed reducing gear so that the motor pulley runs at one fifth of the motor speed. The motor pulley, which is 200 mm in diameter, drives a pulley 300 mm in diameter on the countershaft. A cone pulley on the countershaft drives a similar pulley on the machine tool. Both cone pulleys have the same diameters, viz. 450 mm, 380 mm, 270 mm and 200 mm. Determine the speed range of the machine.

17 The arrangement of a double geared winch is shown in Fig. 11.27 (v). The handle, 400 mm long, is keyed to the driving shaft on which is fastened a 20-tooth pinion A. This gears with wheel B, 60 teeth, at one end of the intermediate shaft. At the other end of the shaft is a pinion C, 25 teeth, which gears with a 65-tooth wheel on the drum shaft. The drum is 500 mm in diameter. If the efficiency of the machine is 70 per cent., what effort applied at the handle will be required to lift a load of 500 N on the drum rope?

18 Fig. 11.27 (vi) shows the gearing diagram for a price computing mechanism. Wheel A is driven at 30 rev/min and gears with B. Wheel B and pinion C are keyed to the same shaft. C gears with two wheels E and D. E is keyed to shaft J. Wheels D and F are keyed to the same shaft and F gears with G keyed on shaft H. The speed of J must be 0·5 rev/min faster than the speed of H. Determine the number of teeth required for wheel G given that the tooth numbers for the other wheels are: A 15, B 75, C 20, D 30, E 40, F 20.

19 Find the mechanical advantage and velocity ratio of a perfectly smooth inclined plane when a force of 28 N parallel to the plane is sufficient to push a weight of 224 N up the plane with uniform velocity. If the surface of the plane became roughened so that the applied force had to be increased to

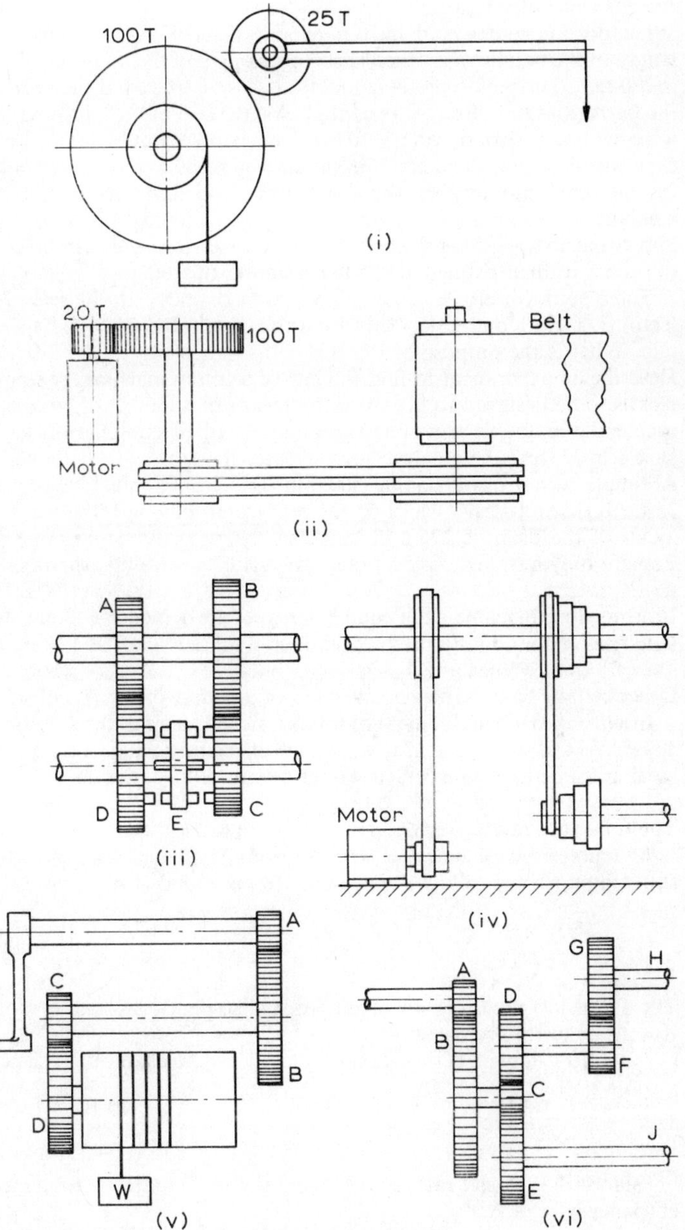
100 T
25 T
(i)
20 T
100 T
Belt
Motor
(ii)
A
B
D
E
C
(iii)
Motor
(iv)
A
C
B
D
W
(v)
G
H
A
D
B
F
C
J
E
(vi)

Figure 11.27

40 N, what would then be the mechanical advantage, the velocity ratio and the efficiency of the plane? [U.L.C.I.]

20 What do you understand by the velocity ratio of a lifting machine? The length of the handle of a winch is 400 mm long and the diameter of the barrel is 200 mm. The pinion on the handle axis has 16 teeth and the spur wheel on the barrel axis has 90 teeth. Find the velocity ratio of the winch. If the winch were without friction, what force at the handle would lift 675 N? In an experiment, a force of 40 N at the handle was necessary to lift 675 N. Account for the difference between the actual force and that necessary in a perfect machine. [U.L.C.I.]

21 Show that if two toothed wheels are in gear their speeds are inversely proportional to their diameters or their numbers of teeth.

Three spur wheels A, B and C on parallel shafts are in gear. A has 12 teeth, B 40 teeth and C 48 teeth. Find the speed of C when A makes 80 rev/min. What is the purpose of wheel B? [U.L.C.I.]

22 Describe an experiment to find the law of a lifting machine. A simple screw jack is to be designed to lift 10 kN by means of an effort of 50 N applied at each end of a double arm of 650 mm total length. Taking the efficiency of the jack as 0·25, find the greatest pitch of screw possible. [N.C.T.E.C.]

23 A simple winch has a handle 300 mm long which rotates a wheel having 35 teeth gearing with a wheel on the drum shaft having 150 teeth. If the diameter of the drum is 225 mm, calculate the velocity ratio of the machine. Explain fully how you would obtain the velocity ratio by experiment. [N.C.T.E.C.]

24 In a hoisting machine, it is found that the effort moves 600 mm while the load rises 40 mm. Find the effort required to lift a load of 10 kN assuming an ideal frictionless machine. [N.C.T.E.C.]

25 Describe how to find the velocity ratio of a simple lifting machine.

In a lifting machine, a weight of 150 N is raised 50 mm when the effort moves 300 mm. Find the mechanical advantage of the machine assuming it to be an ideal machine and state how it would be affected by friction. [N.C.T.E.C.]

26 The following results were obtained from an experiment on a set of pulleys. W denotes the weight raised and E the effort applied. Plot these results on squared paper and obtain the law connecting E and W.

W N	14	21	28	35	42
E N	2·5	3·3	4·25	5·1	6

27 The following figures were taken from a student's laboratory note book. Results of test on screw jack:

Effort applied N	Load lifted N	Mechanical efficiency
0·74	15	0·405
2·5	60	

Deduce the velocity ratio of the machine and fill in the omitted mechanical efficiency. [U.E.I.]

28 A rope lifting tackle with 6 pulleys needs an effort of 16 N to lift a load of 57 N. State the velocity ratio and calculate the efficiency at this load.

29 In a lifting machine, the load moves 10 mm while the effort moves 1·3 m. An effort of 2 N lifts a load of 180 N. Calculate the velocity ratio and the efficiency of the machine.

30 If the efficiency of the machine in question 29 is 50 per cent. when the load is 500 N, calculate the effort required to move this load.

31 A planing machine is driven by a 15 kW motor. The efficiency of the machine is 70 per cent. If the cutting speed is 1 m/s, calculate the cutting force in newtons.

32 A load is raised by means of a pair of pulley blocks, the upper one having 3 pulleys and the lower one having 2. Find (*a*) the velocity ratio, (*b*) the effort required to lift 500 N, and (*c*) the mechanical advantage. The efficiency is 80 per cent. at this load.

33 The larger pulley in a Western pulley block has 11 flats and the smaller one has 10 flats. A load of 200 N is raised by an effort of 23 N. What is the efficiency of the pulley block?

34 From a laboratory test on a system of pulley blocks, a graph of effort against load was prepared, and the following figures were extracted from two points which lay on the graph. The effort was 8·2 N when the load was 20 N, the effort was 20 N when the load was 47 N. The velocity ratio of the machine was 5. What would be the efficiency of the machine when raising a load of 100 N? [N.C.T.E.C.]

35 An ordinary screw jack is operated by a 'tommy bar.' From the design, it is known that an effort of 12 N, applied at a radius of 300 mm, is the ideal effort for a load of 2200 N. What is the lead of the screw?

By experiment, it is found that the actual effort under the above conditions is 13·6 N. What is the efficiency of the screw at this load? [N.C.T.E.C.]

36 A screw jack used for heavy work is manipulated by means of a crank handle of 336 mm crank radius, geared in such a way that two turns of the handle are required to raise the load on the jack 6 mm. Find the velocity ratio, and the load that can be raised by an effort of 80 N applied to the handle if the efficiency is 42 per cent. [N.C.T.E.C.]

37 Give a sketch of a Weston pulley block and describe an experiment to obtain its velocity ratio.

In a block of this type there are 10 flats on the large pulley and 9 on the smaller pulley. What load will be lifted by an effort of 120 N if the efficiency at this load is 40 per cent.?

Why does the weight remain suspended when there is no pull on the lifting chain? [N.C.T.E.C.]

38 What is meant by the 'efficiency of a machine'?

A screw, 3 mm pitch, is tightened up by a spanner 200 mm long. What is the velocity ratio?

If a force of 25 N applied at the end of the spanner causes the screw to exert a pressure of 1200 N, what is the mechanical advantage?

Calculate the efficiency. [N.C.T.E.C.]

39 Describe an experiment to find the law of a lifting machine.

A simple screw jack is to be designed to lift 10 kN by means of an effort of 10 N applied at each end of a double arm of 650 mm total length. Taking the efficiency of the jack as 0·25, find the greatest pitch of screw permissible. [N.C.T.E.C.]

40 In a lifting machine, the velocity ratio is 30 to 1. An effort of 10 N is required to raise a load of 35 N and an effort of 25 N raises a load of 260 N. Find the effort required to raise a load of 165 N and the efficiency at this load. Assume a linear relation between load and effort.

[N.C.T.E.C.]

41 The diagram, Fig. 11.28, illustrates an experimental worm and wheel lifting tackle. Calculate the velocity ratio of the machine, and find what effort would be required to raise a load of 500 N, given that the efficiency at this load is 82 per cent.

Find also the ideal effort for this load. [N.C.T.E.C.]

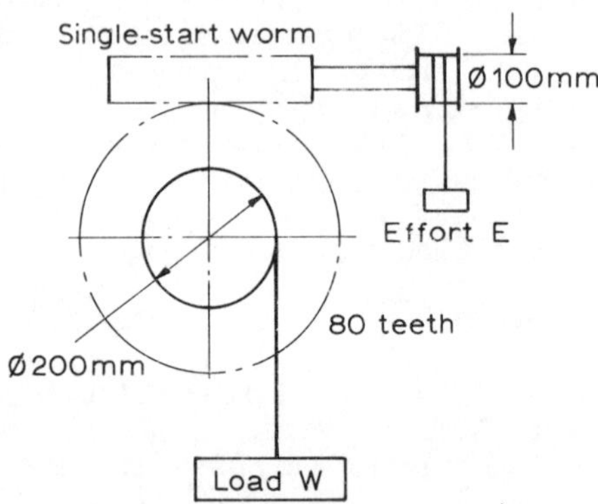

Figure 11.28

42 A screw jack operated by a bar pushed through a hole in the spindle has a single thread of 6 mm pitch. A force of 25 N applied at right angles to the bar at a radius of 350 mm from the axis of the screw, lifts a load of 4000 N. What is the mechanical efficiency of the screw jack at this load?

[N.C.T.E.C.]

43 The following data were obtained for the effort (P) and the load (W) when experiments were made with a small winch: $P = 5{\cdot}7$ N, $W = 40$ N, $P = 19{\cdot}9$ N, $W = 150$ N. The effort-load graph is a straight line and the velocity ratio of the winch is 11. Find the effort required for a load of 200 N and the efficiencies of the winch for loads of 40 N, 150 N, and 200 N. Plot an efficiency curve on a load base.

44 A winding hoist consists of a wheel 1·2 m in diameter and a barrel 300 mm in diameter, mounted on a shaft 75 mm in diameter, and the combination weighs 800 N. If the coefficient of friction between the shaft and the bearings is 0·1, find the vertical downward force P which, acting tangentially to the wheel, is just sufficient to lift a load of 6400 N suspended tangentially from the barrel.

45 A man working at the rate of 40 W is raising a weight of 10 kN by means of an arrangement of lifting tackle. If the velocity ratio is 80 and the efficiency when lifting the load is 70 per cent., what effort is the man exerting and at what rate is the load moving?

46 An experimental screw jack is fitted with a circular load table, the screw being single start, 6 mm in pitch. The effort is applied tangentially to the load table by means of a cord at an effective radius of 200 mm. Calculate the velocity ratio of the machine. If loads of 60 N and 200 N require efforts of 3 N and $4\frac{1}{2}$ N respectively, determine the law of the machine and the efficiencies at the given loads. Why are these two efficiencies different?

[N.C.T.E.C.]

47 A hand-operated winch has two cranked handles, each with an effective radius of 450 mm, which turn a 300 mm diameter drum through a double reduction gear. The first reduction is 4 to 1 and the second 6 to 1. In dragging a piece of machinery into position, the wire rope is loaded to 500 N. If the mechanical efficiency of the winch under this load is 70 per cent., what force is being exerted on each handle? [N.C.T.E.C.]

48 In a differential wheel and axle, the diameter of the effort wheel is 300 mm and the axle diameters are 150 mm and 130 mm. Calculate the velocity ratio. For this machine it was found that an effort of 0·5 N was required to raise a load of 8 N and an effort of 0·9 N was required to raise a load of 18 N. Determine the effort required to raise a load of 15 N and the efficiency at this load. [N.C.T.E.C.]

49 A hoist is driven, through a double reduction gear, by an electric motor which runs at 1000 rev/min. The motor shaft carries a pinion of 15 teeth which engages with a 75-tooth gearwheel keyed to a shaft which also carries a 14-tooth pinion. This latter engages with a 98-tooth gearwheel keyed to the hoist drum shaft. If the effective diameter of the hoist drum and rope is 200 mm and the load on the hoist rope is 20 kN, determine: (*a*) the speed at which the load is being raised, (*b*) the power exerted by the motor if the gear efficiency is 80 per cent. Neglect the weight of the rope.

[N.C.T.E.C.]

Chapter Twelve
Centre of Gravity

Let us think of any solid (a casting for example) as being made up of a very large number of particles. Each of these particles is attracted towards the centre of the earth by the force of gravity. A very large number of small forces are consequently acting upon the solid. All these forces will be

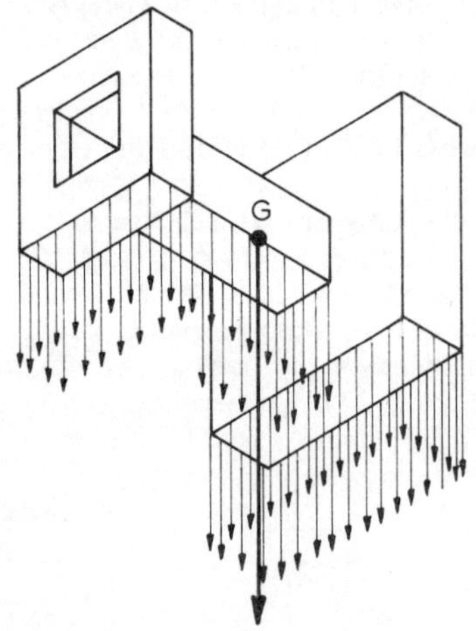

Figure 12.1 *Centre of Gravity*
The pull of the earth acts on every particle in a body. All these forces are parallel. The resultant of these forces passes through a point known as the centre of gravity.

parallel to each other (actually, if you produce the lines of action of all these forces they will meet at the centre of the earth, but this is such a long way away that we can think of the forces as being parallel). The resultant of all the forces will be equal to the weight of the solid. The point through which this resultant will act is called the **centre of gravity.**

Definition. The centre of gravity of a body is that point through which the resultant gravitational force acts in whatever position the body may be placed.

The position of the centre of gravity (c.g.) of a number of solids can be determined by inspection.

For example:

The c.g. of a cube is at the centre of the cube or at the intersection of the cross diagonals.

The c.g. of a sphere is at the centre of sphere.

The c.g. of a flat disc is at the centre of the disc.

The c.g. of a rod of uniform diameter is at the mid-point of its length.

It is not possible, however, to determine the position of the centre of gravity of all solids in this manner. To find the position of the centre of gravity of a pyramid or cone, we shall have to resort to mathematical methods later on, while to find the position of the centre of gravity of an irregular solid we shall have to apply experimental methods.

If we can divide a compound solid up into a number of simpler solids, the position of whose centre of gravity is known, then we can determine the position of the centre of gravity of the whole solid. The calculation depends on the principle that:

> *the sum of the moments of all the parts about any line is equal to the moment of the whole solid about the same line.*

In Fig. 12.2 we are given details of a solid in the form of a T-square. We require to find the position of the centre of gravity of the solid.

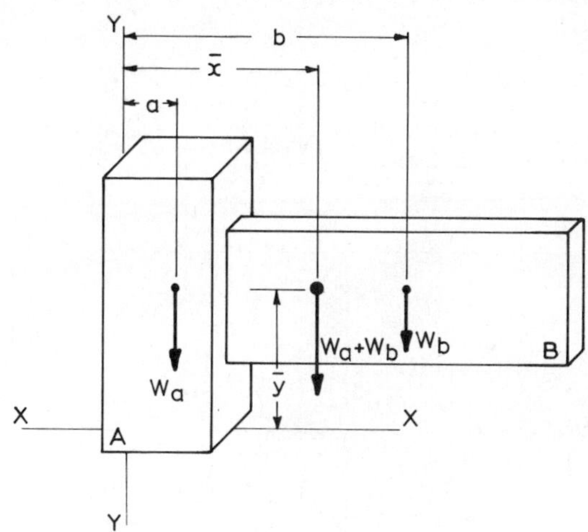

Figure 12.2

Method

(1) Divide the solid into two blocks, A and B, whose weights are W_a and W_b respectively.

(2) Indicate the position of the centre of gravity of each part.

(3) Choose a line YY about which we can take moments. This line will usually be at one end of the solid. Calculate the distance of the centre of gravity of A and B from this line. These distances are indicated as *a* and *b*.

(4) Calculate the moment of each part about YY and add these moments together.

(5) If the distance of the centre of gravity of the whole solid from YY is indicated by $\bar{x}$ (called 'bar *x*') and its total weight is W, then the moment of the whole solid is W$\bar{x}$ about the YY line. This moment should equal the sum of the moments as given in stage 4:

$$W\bar{x} = W_a a + W_b b$$

$$\bar{x} = \frac{W_a a + W_b b}{W}$$

(6) The calculation is then repeated, working from another line XX at right angles to YY. In this way we obtain the distance of the centre of gravity from the base or the distance $\bar{y}$.

Tabular solution

If the solid is complicated, it is often advisable to set the calculations out in tabular form, using the following table:

Part	Weight	Distance of c.g. from XX	Moment about XX	Distance of c.g. from YY	Moment about YY
Total weight		Sum of moments about XX		Sum of moments about YY	

$$\bar{x} = \frac{\text{sum of moments about YY}}{\text{total weight}}$$

$$\bar{y} = \frac{\text{sum of moments about XX}}{\text{total weight}}$$

Short cuts

There are a number of ways in which the calculations can be reduced in certain cases:

(1) A symmetrical figure, i.e. one in which we can find a line, so that the figure on one side is a reflection of the figure on the other side.

For example, a circle is symmetrical about a diameter; an isosceles triangle is symmetrical about a line bisecting the angle between the equal sides; a square about a diagonal or about a line joining the mid-points of opposite sides.

In all these cases the centre of gravity is on the line of symmetry.

(2) If the solid is made up of the *same material throughout*, we can deal with *volumes instead of weights*, since the weight of each part is simply density × volume and the density will be the same for each part.

(3) If the solid is (*a*) made from the *same material*, and (*b*) is the *same thickness* throughout, we need deal only with the *face area of each part*, since the weight of each part is

face area × thickness × density.

The product (thickness × density) is the same for each part.

(4) If the solid is made of bent wire, the *diameter and material being uniform* throughout, then we need deal only with the *length of each part*, since the weight of each part is

length × cross-sectional area × density.

The product (cross-section area × density) is the same for each part.

Centroid

We use the term 'centre of gravity' when dealing with a solid which has mass.

The term centre of area or 'centroid' is used to denote the corresponding point in connection with a plane area or section which cannot have mass. Thus we speak of the centroid of a square, but the centre of gravity of a cube.

The method of determining the position of the centroid is the same as that used to find the position of the centre of gravity, with the exception that areas will be used in place of masses.

The position of the centroid of four of the most important sections is shown in Fig. 12.3.

(1) The centroid of a narrow strip is at its mid-point.

(2) The centroid of a rectangular area is at the intersection of the lines joining the mid-points of opposite sides. It is also at the point of intersection of the diagonals.

(3) The centroid of a circular area is at the geometrical centre.

(4) The centroid of a triangular area is at the point of intersection of the medians or lines joining the apex of the triangle to the mid-point of the opposite side. The distance of this point from any base is equal to one-third of the corresponding height of the triangle.

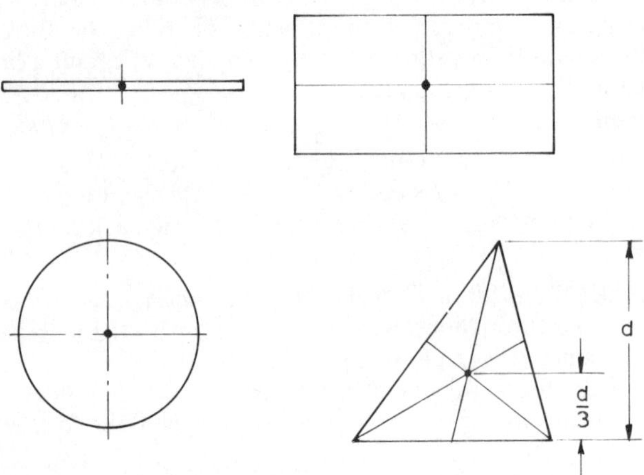

Figure 12.3 *Centroids of Simple Sections*

Whilst the position of the centroid in the first three of the above sections may be clear, it may be necessary to indicate the reason for the statement concerning the triangle. Consider the triangle shown in Fig. 12.4 as being

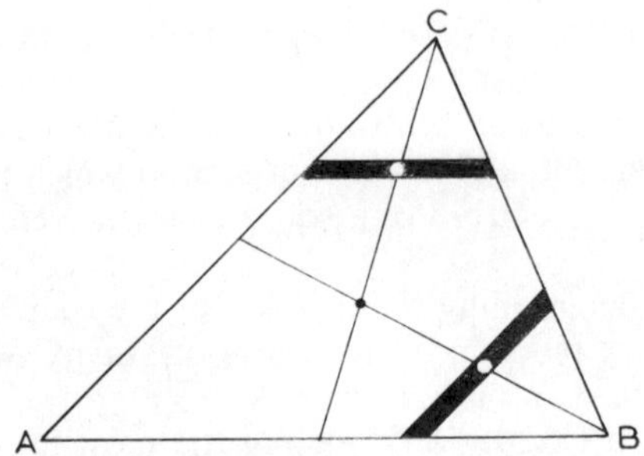

Figure 12.4 *Position of Centroid of Triangle*
The centroid of a triangle is at the inter-section of the lines joining the corner of the triangle to the mid-point of the opposite side.

made up of a number of strips parallel to the base AB. One such strip is shown. The centroid of the strip is at its mid-point. The centroids of all these parallel strips will lie on the line joining C to the mid-point of the base AB. Therefore, the centroid of the whole area must lie on this line. Similarly, consider the area to be made up of a number of strips parallel to AC, one of which is shown. The centroid of this strip is at its mid-point. The centroids of all these parallel strips lie on the line joining B to the mid-point of AC. Therefore, the centroid of the whole area must be on this line. That is, the centroid of the triangular area must be at the intersection of the lines joining the apex of the triangle to the mid-point of the opposite side.

Application of the centre of gravity

Think of any solid whose weight is, say, W and whose centre of gravity is at G. The solid is suspended from a point O. Two forces are acting on the solid:

(1) Its own weight, W, vertically down through G.
(2) The upward pull in the cord, which will also equal W.

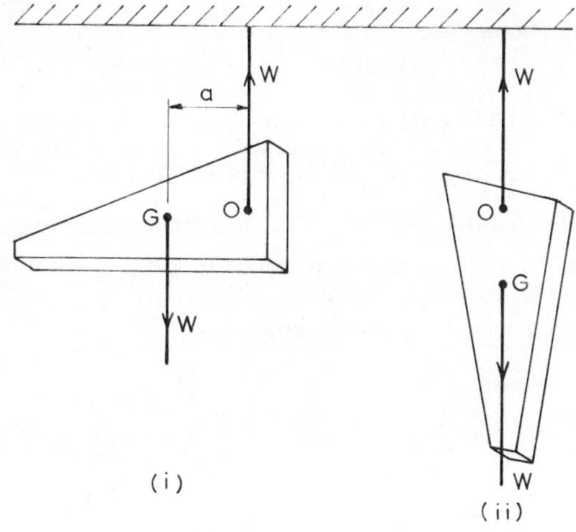

Figure 12.5 *Point of Support and Centre of Gravity*
When a solid is hanging from a point of support, the centre of gravity will lie on a vertical line through the supporting point.

Now, if the solid is hanging as shown in Fig. 12.5(i), it obviously will not stay in that position if it is free to swing about O. The two forces form a couple whose magnitude is Wa and whose direction is anticlockwise. The solid will therefore turn in an anticlockwise direction. As it turns, the distance a will decrease, and therefore the magnitude of the couple will decrease. When G is vertically below O, as shown in Fig. 12·5(ii), the magnitude of the couple is reduced to zero and the solid is in equilibrium with G on a vertical line through O, the point of support.

This fact is true of all suspended solids. *The centre of gravity is always on a vertical line through the point of suspension.* This is why the position of the centre of gravity of a loaded crate is marked on the outside of the crate so that it can be slung safely by a crane.

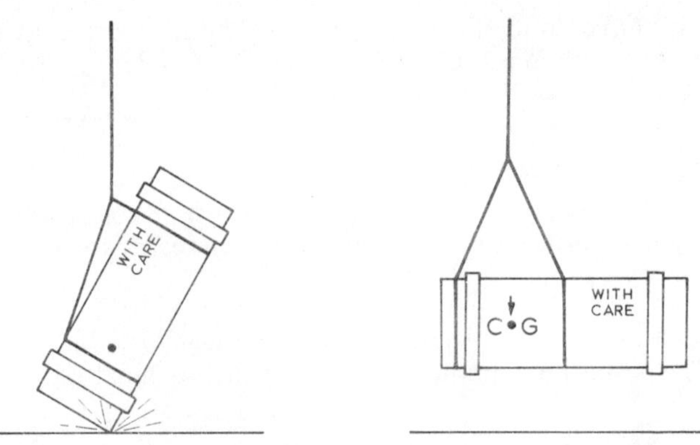

Figure 12.6 *Importance of Knowing Position of Centre of Gravity*
The centre of gravity is not always at the geometrical centre of the solid. In dealing with heavy crates it is necessary to know the position of the centre of gravity to avoid accidents when lifting. It is usual in such cases to indicate the position of the centre of gravity by a suitable mark on the case. The case must then be lifted so that the centre of gravity lies on the line of pull of the crane.

Now imagine that the solid is allowed to rest on a horizontal surface instead of being suspended from a point O. If the solid is placed as shown in Fig. 12·7(i), the weight acting through G will produce an anticlockwise

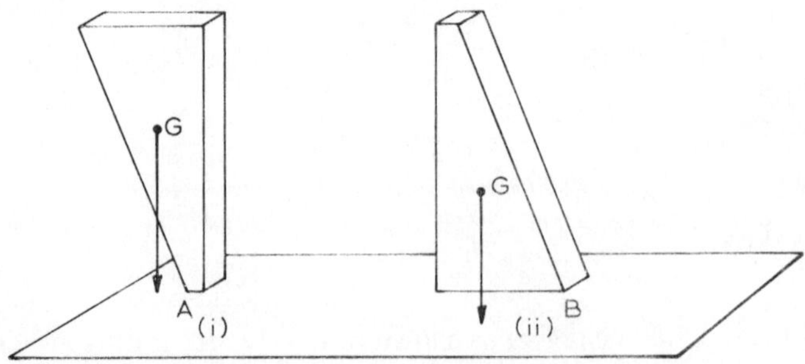

Figure 12.7 *Position of Centre of Gravity and the Supporting Base of the Solid*
A solid will only remain in a position of equilibrium if a vertical line through the centre of gravity passes through the base of the solid.

turning moment about A and the solid will topple over in an anticlockwise direction. If the solid is placed as shown in Fig. 12·7(ii), then the weight produces an anticlockwise moment about B, but the solid obviously cannot turn because the surface prevents its movement and the solid is consequently in equilibrium.

Notice particularly the difference between the positions of G.

In the first case, a vertical line through G passes *outside the base of the solid*, and the *solid is not in equilibrium*. In the second case a vertical line through G *cuts the base of the solid*, and the *solid is in equilibrium*.

A body resting upon any surface will be in equilibrium if a vertical line through the centre of gravity passes through the base of the solid. (Notice that the body is stated to be at rest; when we think of rotating bodies, such as a cyclist moving round a circular track, we may find that a vertical line through the centre of gravity will not pass through the base. This is because effects other than gravity are acting on the body.)

Stable and unstable equilibrium

If you are very patient you may be able to balance a cone upon its point. When you have managed that, just move it slightly from its equilibrium position. Then take your hand away. Will the cone return to its original position of balance? No, it will not; it will fall over into a new position of rest.

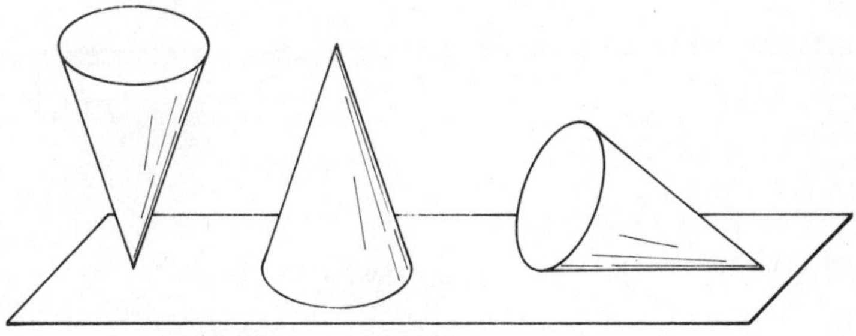

Figure 12.8 *Unstable, Stable and Neutral Equilibrium*

The three cones are all in equilibrium, although the first one is only just. If these cones are slightly displaced from the positions shown, the first will fall over into a new position, the second will return into its original position and the third will stay in its displaced position. The first is said to be in unstable equilibrium, the second in stable and the third in neutral equilibrium.

Now stand the cone on its base. Then tilt it over just a little and take your hand away. The cone will fall back into its original position.

Finally, let the cone rest along the sloping side so that the apex is on the table and the base is inclined. Now just roll it from this position and take your hand away. If it is a perfect cone it will neither return to its old position nor move into a new position. It will just 'stay put'.

The first position is a position of unstable equilibrium.

The second position is a position of stable equilibrium.

The third position is a position of neutral equilibrium.

A body is in *stable equilibrium* if, when slightly displaced from its position, it will *return again to that position*.

A body is in *unstable equilibrium* if, when slightly displaced from its position, it will *move into another position* of equilibrium.

A body is in *neutral equilibrium* if, when slightly displaced from its position, it will *remain in the displaced position.*

Example 12.1

Details of a gear forging are given in Fig. 12.9. Determine the position of the centre of gravity if the density of the metal is 7830 kg/m³ ($7{\cdot}83 \times 10^{-6}$ kg/mm³).

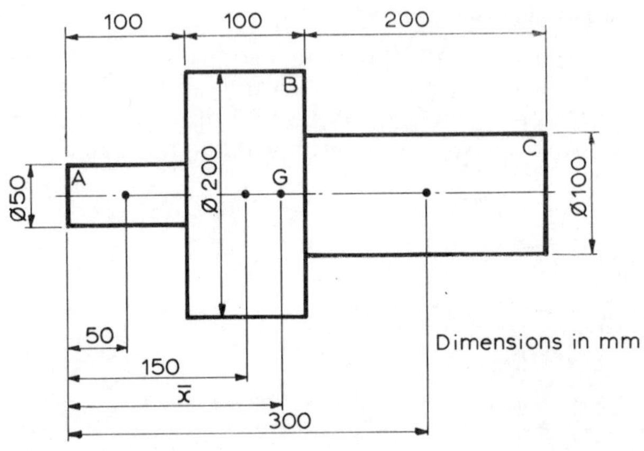

Figure 12.9

The forging is considered as made up of the three parts A, B and C, the centre of gravity of each part being on the central axis at the mid-point of each portion.

$$\text{Weight of A} = \frac{\pi}{4} \times 50^2 \times 100 \times 7{\cdot}83 \times 10^{-6}g = 1{\cdot}54g \text{ N}$$

$$\text{Weight of B} = \frac{\pi}{4} \times 200^2 \times 100 \times 7{\cdot}83 \times 10^{-6}g = 24{\cdot}6g \text{ N}$$

$$\text{Weight of C} = \frac{\pi}{4} \times 100^2 \times 200 \times 7{\cdot}83 \times 10^{-6}g = 12{\cdot}3g \text{ N}$$

$$\text{Total weight} = 38{\cdot}44g \text{ N.}$$

Distance of c.g. of A from left-hand end

$$= \frac{100}{2} = 50 \text{ mm.}$$

Distance of c.g. of B from left-hand end

$$= 100 + \frac{100}{2} = 150 \text{ mm.}$$

Distance of c.g. of C from left-hand end

$$= 100 + 100 + \frac{200}{2} = 300 \text{ mm.}$$

Taking moments about left-hand end:

$$\begin{aligned} \text{Moment of A} &= 1{\cdot}54g \times 0{\cdot}050 = 0{\cdot}077g \text{ Nm} \\ \text{Moment of B} &= 24{\cdot}6g \times 0{\cdot}150 = 3{\cdot}690g \text{ Nm} \\ \text{Moment of C} &= 12{\cdot}3g \times 0{\cdot}300 = 3{\cdot}690g \text{ Nm} \end{aligned}$$

$$\text{Sum of moments} = 7{\cdot}457g \text{ Nm.}$$

$$\text{Total weight} \times \bar{x} = \text{sum of moments}$$

$$38{\cdot}44\, g\bar{x} = 7{\cdot}457g$$

$$\bar{x} = \frac{7{\cdot}457}{38{\cdot}44} = 0{\cdot}194 \text{ m} = 194 \text{ mm.}$$

Centre of gravity is on axis of forging at a distance of 194 mm from the left-hand end.

Example 12.2

Determine the position of the centroid of the angle-iron section shown in Fig. 12.10.

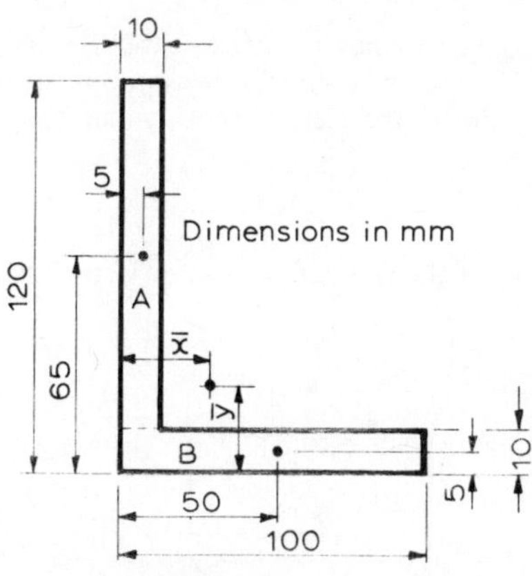

Figure 12.10

The section is divided into 2 rectangles as shown. The centroid of each section is at the mid-point of the rectangle.

$$\begin{aligned} \text{Area of A} &= 10(120 - 10) = 1100 \\ \text{Area of B} &= 10 \times 100 \quad\;\; = 1000 \end{aligned}$$

$$\text{Total area} = 2100$$

To find $\bar{x}$, take moments about left-hand edge:

$$\text{Moment of A} = 1100 \times 5 = 5500$$
$$\text{Moment of B} = 1000 \times 50 = 50\,000$$
$$\text{Total moment} = 55\,500$$

$$\text{Moment of whole area} = \text{sum of moments of parts}$$
$$2100\,\bar{x} = 55\,500$$
$$\bar{x} = 26{\cdot}4 \text{ mm.}$$

To find $\bar{y}$, take moments about base.

Note carefully that the distance of centroid for part A from the base is $10 + \frac{110}{2} = 65$ mm. This often causes difficulty and leads to incorrect answers.

$$\text{Moment of A} = 1100 \times 65 = 71\,500$$
$$\text{Moment of B} = 1000 \times 5 = 5000$$
$$\text{Total moment} = 76\,500$$

$$\text{Moment of whole area} = \text{sum of moments of parts}$$
$$2100\,\bar{y} = 76\,500$$
$$\bar{y} = \frac{76\,500}{2100} = 36{\cdot}4 \text{ mm.}$$

The centroid of the section is 26·4 mm from the left-hand edge and 36·4 mm from the base.

Note that the centroid is outside the area of the figure. This is often the case with angle sections. If you were to cut the section out of cardboard, you would not be able to find a point in the section about which the section would be balanced for all positions.

Example 12.3

Determine the position of the centroid of the section shown in Fig 12.11.

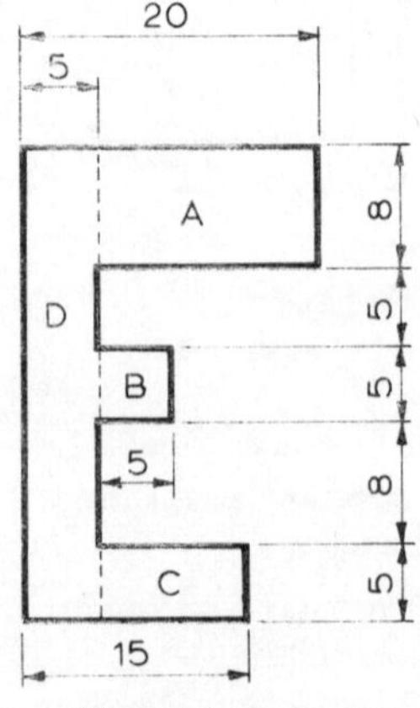

Figure 12.11

The section is divided into four rectangles, the centroids of which are at the mid-point of each rectangle.

This example will be solved by the tabular method.

Part	Area cm^2	Distance from l.-h. edge	Moment about l.-h. edge	Distance from base	Moment about base
A	120	12·5	1500	27	3240
B	25	7·5	187·5	15·5	387·5
C	50	10	500	2·5	125
D	155	2·5	387·5	10·5	1627·5
Total area	350	Sum of moments about l.-h. edge	2575	Sum of moments about base	5380

$$\bar{x} = \frac{\text{sum of moments about l.-h. edge}}{\text{total area}} = \frac{2575}{350} = 7{\cdot}36 \text{ cm}$$

$$\bar{y} = \frac{\text{sum of moments about base}}{\text{total area}} = \frac{5380}{350} = 15{\cdot}37 \text{ cm.}$$

The centroid of the section is 7·36 cm from the left-hand edge and 15·37 cm from the base.

Example 12.4

A plate weighing 15 N has the shape and dimensions given in Fig. 12·12. It

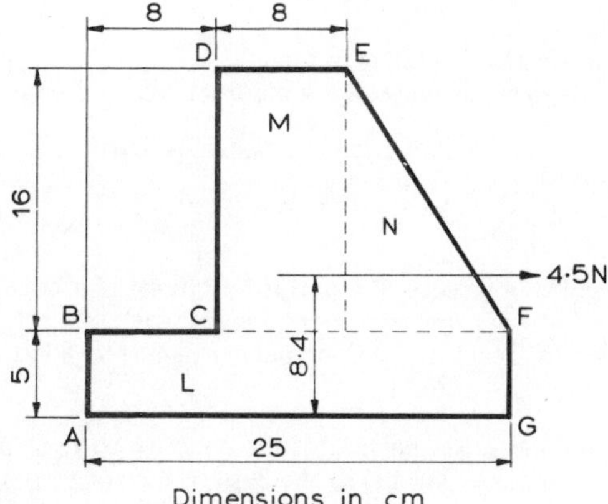

Figure 12.12

rests with its face on a horizontal surface, the coefficient of friction, μ, being 0·3. Determine the magnitude, direction and position of the least force which will slide the plate in a direction parallel to AG without rotating it. [U.L.C.I.]

The magnitude of the force to move a plate whose weight is 15 N over a surface where μ is 0·3 is $15\mu = 4{\cdot}5$ N.

If the plate is to move in a direction parallel to AG, then the force must be applied in a direction parallel to AG. If the plate has to move without turning, then the force must be applied along a line passing through the centre of gravity of the plate.

Position of c.g.

Take moments about AG, working with areas only.

Part	Area	Distance of c.g. from AG	Moment about AG
L	$5 \times 25 = 125$	2·5 cm	$125 \times 2{\cdot}5 = 312{\cdot}5$
M	$8 \times 16 = 128$	$5 + [\frac{1}{2} \times 16] = 13$ cm	$128 \times 13 = 1664$
N	$\frac{1}{2} \times 9 \times 16 = 72$	$5 + [\frac{1}{3} \times 16] = 10{\cdot}33$ cm	$72 \times 10{\cdot}33 = 743{\cdot}8$
	Total area = 325		Total moment = 2720·3

If c.g. is $\bar{y}$ from AG,

$$\text{then } 325\bar{y} = 2720{\cdot}3$$
$$\bar{y} = 8{\cdot}4 \text{ cm.}$$

Centre of gravity is 8·4 cm from AG.

To move the plate without rotating, a force of 4·5 N must be applied in direction parallel to AG acting at a distance of 8·4 cm from AG as shown.

(Note in this question we only need the distance of the c.g. from the base AG. Its distance from AB is not required.)

Example 12.5

Given that the centre of gravity of a uniform right circular cone is at a distance from the base equal to a quarter of the height, calculate the position of the centre of gravity of a frustum of a cone having diameters 8 cm and 4 cm and a height of 5 cm.

A frustum of a cone is the portion of the cone which remains when the top part is cut away in a plane parallel to the base of the cone.

Let us think of the portion which was removed as now being replaced and shown dotted in Fig. 12.13.

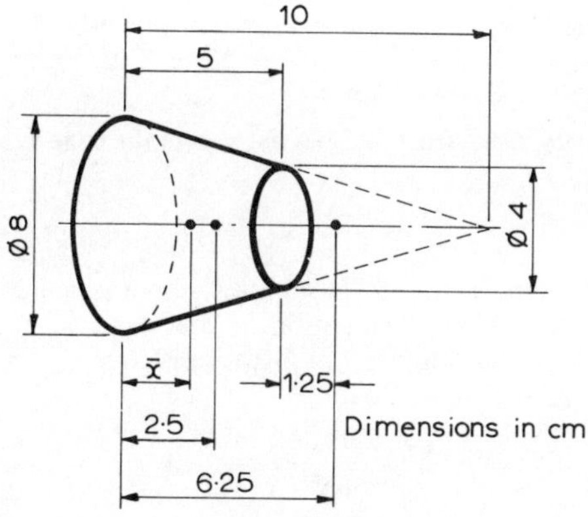

Figure 12.13

Let L = total height of cone.

Then by similar triangles $\dfrac{L}{8} = \dfrac{L-5}{4}$

which gives L = 10 cm.

Volume of large cone $= \frac{1}{3} \times \dfrac{\pi}{4} \times 8^2 \times 10$ ($\frac{1}{3}$ base area $\times$ height)

$= 168$ cm³.

Distance of c.g. from base $= \dfrac{\text{height}}{4} = \dfrac{10}{4}$

$= 2{\cdot}5$ cm.

Now the whole cone is made up of two portions:

(1) The small cone:

Height = 5 cm.

Volume $= \frac{1}{3} \times \dfrac{\pi}{4} \times 4^2 \times 5 = 21$ cm³.

Distance of c.g. from its own base

$= \dfrac{\text{height}}{4} = 1{\cdot}25$ cm

Distance of this c.g. from base of whole cone

$= 1{\cdot}25 + 5$

$= 6{\cdot}25$ cm.

(2) The frustum:

Volume = volume of whole cone − volume of small cone.

$= 168 - 21$

$= 147$ cm.

Distance of c.g. from base of frustum is not known.

Let this length be $\bar{x}$.

Taking moments about the base of the frustum:

Moment of whole cone = moment of frustum + moment of small cone.

$$168 \times 2{\cdot}5 = 147\bar{x} + 21 \times 6{\cdot}25$$
$$\bar{x} = 1{\cdot}97 \text{ cm.}$$

The centre of gravity of the frustum is on the axis of the cone at a distance 1·97 cm from the base.

Example 12.6

A cylinder 3 cm in diameter and 5 cm high rests on an inclined plane in such a way that it cannot slip. If the inclination of the plane is gradually increased, find the angle at which the cylinder will just topple over.

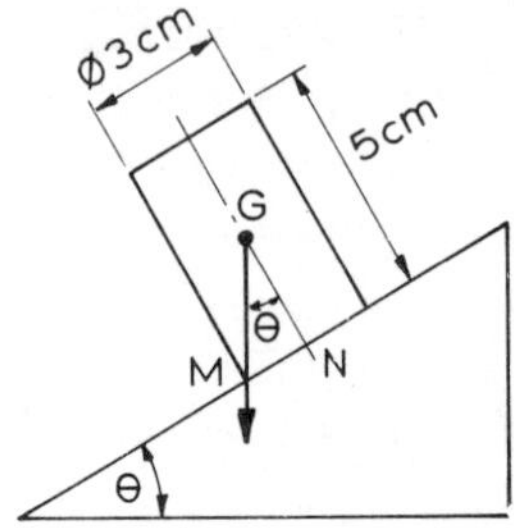

Figure 12.14

The cylinder will be in equilibrium and will not topple over *provided that a vertical line through the centre of gravity G falls within the base.*

The limiting position is when the vertical passes the point M as shown in Fig. 12.14.

The angle of the plane θ which is also the angle NGM can be calculated or measured from a scale drawing. The centre of gravity of the cylinder will be on the axis at distance $\frac{1}{2} \times 5$ from the base.

$$\therefore \quad GN = 2{\cdot}5 \text{ cm}$$
$$MN = 1{\cdot}5 \text{ cm}$$
$$\tan\theta = \frac{MN}{GN} = \frac{1{\cdot}5}{2{\cdot}5} = 0{\cdot}6$$
$$\theta = 30^\circ\ 58'$$

The cylinder will topple over if the angle of the plane is greater than 30° 58′.

EXERCISE 12

1–8 Determine the position of the centroid in each of the eight sections given in Fig. 12.15. In each case, give the distance of the centroid from the farthest edge on the left-hand side and also from the lowest edge or base of the section.

9 A piece of wood, 4 cm square by 30 cm long, is cut into two pieces, 10 cm and 20 cm long respectively, and made into a T by gluing the 10 cm piece to the end of the 20 cm piece. Find the position of the centre of gravity of the solid thus formed.

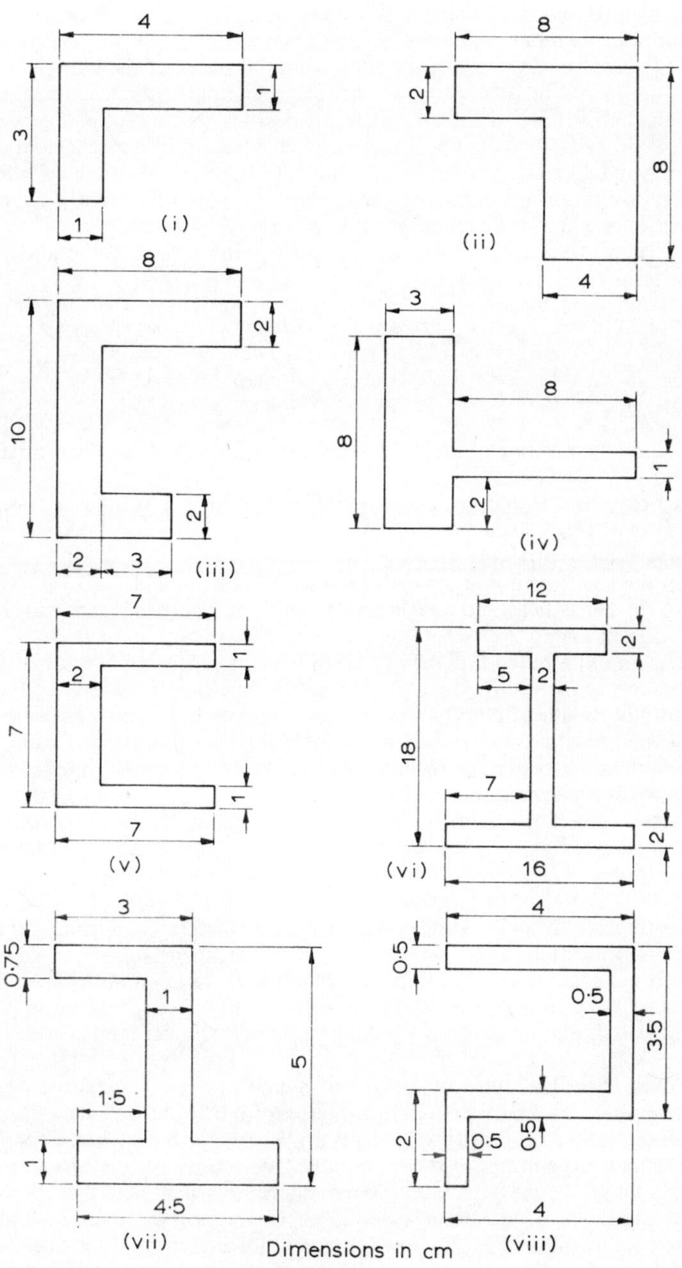
4
1
3
1
(i)
8
2
8
4
(ii)
8
2
10
2
2
3
(iii)
3
8
8
1
2
(iv)
7
1
2
7
1
7
(v)
12
2
5
2
18
7
2
16
(vi)
3
0·75
1
5
1·5
1
4·5
(vii)
4
0·5
0·5
3·5
2
0·5
0·5
4
(viii)
Dimensions in cm

Figure 12.15

10 A piece of wood of uniform thickness consists of a square of side 200 mm surmounted by an isosceles triangle of height 150 mm. Calculate the distance of the centre of gravity of the solid from the vertex of the triangle.

11 A uniform wire 300 mm long is bent at right angles, the lengths of the arms being in the ratio of 3:1. Calculate the position of the centre of gravity.

12 A solid body is made up of a cylinder of 150 mm in diameter and 200 mm long and a hemisphere of 75 mm radius fitting one end of the cylinder. Find the position of the centre of gravity of the whole solid body, given that the centre of gravity of a hemisphere is $\frac{3}{8}$ × radius from the base.

13 A bar of metal 60 cm long has been turned down to three parallel portions as follows:

	length	diameter	weight
A	15 cm	6 cm	37 N
B	20 cm	5 cm	32 N
C	25 cm	4 cm	22 N

Find the position of the centre of gravity of the bar from the 4 cm diameter end.

14 A heavy bar of uniform cross-section is 4 m long. Find the position of the centre of gravity when:
(*a*) the bar is straight.
(*b*) the bar is bent into a right angle, each leg 2 m long.
(*c*) the bar is bent into a right angle, with one leg 2·5 m long, and the other 1·5 m long.

15 The arm of a radial drill may be considered as consisting of a solid of weight 2·5 kN, whose centre of gravity is 1 m from the end of the arm, and a second solid of weight of 2 kN whose centre of gravity is 1·8 m from the same end of the arm. It is to be lifted by a single rope sling suspended from a crane hook. At what distance from the end of the arm must the rope be placed so as to lift the arm without tilting it?

16 A telescope consists of three tubes, 100, 150 and 200 cm long respectively, and their weights, in the same order, are 1·5, 2 and 2·5 N. Determine the position of the centre of gravity of the telescope when fully extended, giving its distance from the large end.

17 Distinguish between stable, unstable and neutral equilibrium, giving an example of each.

A cylinder, 10 cm diameter and 20 cm high, rests on an inclined plane in such a way that it cannot slip. If the inclination of the plane is gradually increased, find an angle at which the cylinder will just topple over. [N.C.T.E.C.]

18 Given that the centre of gravity of a uniform right circular cone is at a distance from the base equal to a quarter of the height, calculate the position of the centre of gravity of a frustum of a cone having radii 75 mm and 125 mm if the height of the frustum is 250 mm. [N.C.T.E.C.]

19 A right solid cone has a base 100 mm diameter, a slant height of 100 mm, and weighs 20 N. It stands on a level bench, and is then tilted 30° out of the vertical, without slipping, by a horizontal force applied at the apex. Find (*a*) the necessary force to maintain the cone in this position, (*b*) the reaction of the ground in magnitude and direction. A graphic or a calculated solution may be offered.

NOTE. The centre of gravity of a right solid cone lies on its axis, one-quarter of the height measured from the base.

[N.C.T.E.C.]

20 The outside dimensions of an angle-iron section are 100 mm by 50 mm and it is 10 mm thick. Regarding all corners as sharp, find the position of the centre of gravity of a thin section of this angle-iron, stating its distances from the 100 mm and 50 mm outside edges. [N.C.T.E.C.]

21 A metal plate with a rectangular opening is shown in Fig. 12.16. The thickness is everywhere the same. Find the position of the centroid of the sheet.

[N.C.T.E.C.]

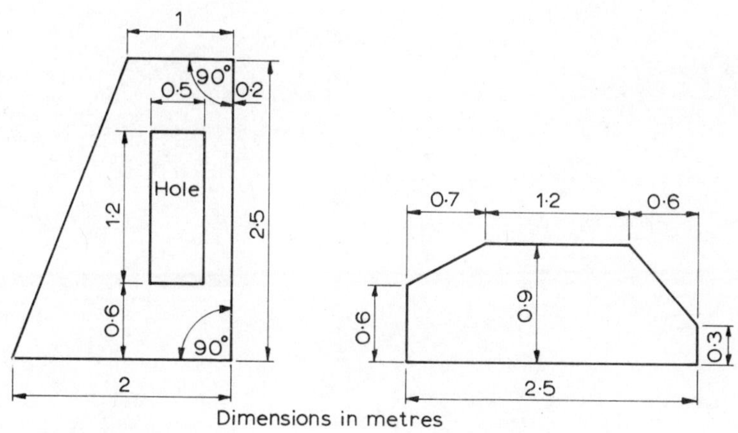

Figure 12.16 **Figure 12.17**

22 Find the position of the centroid of surface area of an irregular-shaped portion of a sheet of metal of uniform thickness shown in Fig. 12.17.

[N.C.T.E.C.]

23 A steel bar, circular in section, 20 cm overall length, is 2 cm in diameter for a length of 10 cm, and tapers to a point in the length of the remaining 10 cm. Find the position of the centre of gravity. [N.C.T.E.C.]

24 Define the centre of gravity of a body.

A circular disc of metal of uniform thickness has a hole 5 cm in diameter cut through it. The distance between the centre of the hole and the centre of the disc is 25 cm. Find what sized hole, with its cente 15 cm from the centre of the disc, would just balance the disc for rotation [N.C.T.E.C.]

25 A brick wall 3·6 m high has its back vertical and is 1 m wide at the base and 0·5 m wide at the top. There are two intermediate steps of 25 cm at 1·2 m centres in the height of the wall. Calculate the distance of its centre of gravity from the back and from the base. [N.C.T.E.C.]

26 Define centre of gravity.

A rectangular box of uniform thickness, complete with top, weighs 450 N; the external dimensions are: length 2 m, width 1·5 m, height 1·2 m. A weight of 100 N is fixed to the top of the box in such a position that it may be considered as being concentrated at a point on the top face of the box 1 m from each end and 40 cm from one of the top long edges of the box. If the top face of the box is horizontal, determine the horizontal distance between the centres of gravity of the box and weight together, and the box alone.

[N.C.T.E.C.]

Chapter Thirteen
Force, Friction and Funicular

When a body at rest is in equilibrium under the action of a number of co-planar forces (i.e. forces which are all acting in one plane), two important facts can be deduced:

(1) There is no tendency for the body to move in any direction; there is, therefore, *no resultant force acting on the body*, i.e. the force diagram is a closed figure.
(2) There is no tendency for the body to rotate; there is, therefore, *no unbalanced moment about any point*.

These two important conditions for equilibrium apply to all cases where co-planar forces acting on a body produce equilibrium.

In addition, there are two ideas which can be applied to the special cases in which:

(*a*) two forces, and
(*b*) three forces act on a body to produce equilibrium.

Two Forces. The two forces must be equal and opposite and they must act in the same straight line.

Three Forces. The lines of action of the **three forces must all pass through one point,** i.e. the three forces must be concurrent. The only exception to this rule is in the case of three parallel forces acting on the body.

The general conditions for the equilibrium of a body can be stated as follows:

(1) The vector diagram drawn for the forces must be a closed diagram.
(2) The algebraic sum of the moments of all the forces about any point must be zero.

In the case of three forces acting on the body, this second condition also implies that the line of action of the three forces must pass through one point.

Direction of reaction

If a beam or a framework rests upon rollers, or is defined as being simply supported, both these conditions imply that the beam or the framework rests upon smooth supports, and consequently the line of action of the reaction is at right angles to the supporting surface.

If the beam or the framework is hinged, then the line of action of the hinge can be in any direction, such direction being determined by the loading conditions.

Resolution of forces

Many problems in statics can be solved easily by applying ideas connected with the resolution of forces. If a problem is one which requires the resultant of a number of forces to be found, then we can use the principle that the component of the resultant of the forces in any direction is equal to the algebraic sum of the components taken in the same direction, of all the forces.

If the problem is one in which it is stated that the body is in equilibrium under the action of a number of forces, then we can use the idea that since the body is in equilibrium, there will be no resultant component force acting in any direction.

Example 13.1

ABCD is a square of $2\frac{1}{2}$ m side. A force of 4 N acts from A to B, of 6 N from C to B, of 7 N from C to D, and of 2 N from A to D. Determine the resultant of these forces and state its position measured from D.

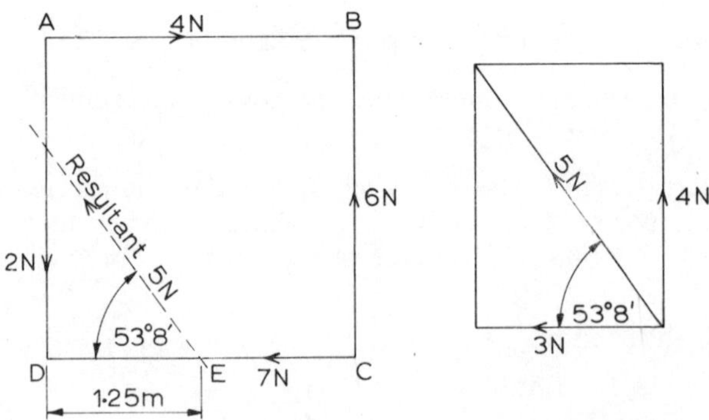

Figure 13.1 *Resultant of Forces*

The horizontal and vertical components of the resultant force will equal the algebraic sum of the respective horizontal and vertical components of the forces acting on the square.

Resolve horizontally: To the left considered positive

$$7 - 4 = 3 \text{ N to the left}$$

Forces in AD and CB have no horizontal component.

Resolve vertically: Upwards considered positive

$$6 - 2 = 4 \text{ N upward}$$

Forces in AB and CD have no vertical component. The 3 N and 4 N components are now combined vectorially, giving the resultant force equal to 5 N inclined at 53° 8′ to the horizontal. The position of the force is obtained by realising that the moment of the resultant about any point is equal to the algebraic sum of the moments of the forces. Take moments about any point, say D. Forces AD and CD have no moment about D since they pass through this point.

Moments about D (anticlockwise considered positive)

$$= -(4 \times 2\tfrac{1}{2}) + (6 \times 2\tfrac{1}{2})$$
$$= 5 \text{ Nm anticlockwise.}$$

Resultant must be placed so that its moment about D will be 5 Nm anticlockwise, i.e. at 1 m from D (perpendicular distance).

Let the line of action of the resultant cut DC in E.

$$\text{Then DE} = \frac{1}{\sin 53^\circ 8'} = \frac{1}{\frac{4}{5}} = 1{\cdot}25 \text{ m}$$

Resultant is 5 N inclined at 53° 8′ to DC as shown, cutting DC at 1·25 m from D.

Example 13.2

Determine the resultant of the following forces which are all acting away from the same point. The angles are measured from a horizontal line in an anti-clockwise direction:

(1) 10 N at 60°
(2) 20 N at 150°
(3) 15 N at 225°
(4) 5 N at 300°

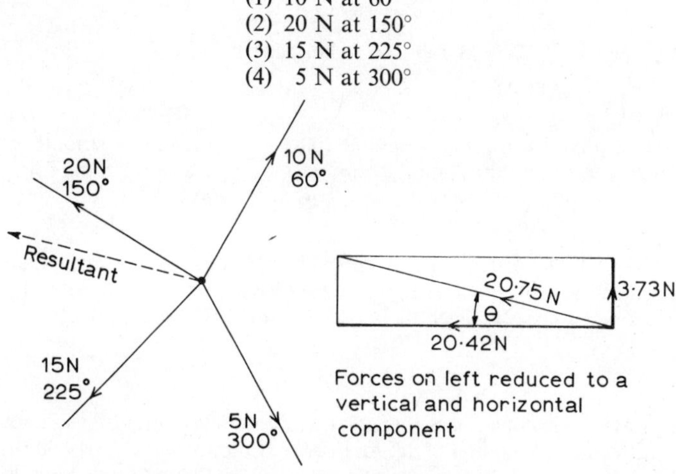

Figure 13.2 *Resultant of Forces*

The horizontal and vertical components of the resultant are equal to the sum of the horizontal and vertical components respectively of the forces.

Resolve horizontally: Horizontal component $= P \cos \theta$.

Sum of components

$$= 10 \cos 60^\circ + 20 \cos 150^\circ + 15 \cos 225^\circ + 5 \cos 300^\circ$$
$$= 10 \cos 60^\circ - 20 \cos 30^\circ - 15 \cos 45^\circ + 5 \cos 60^\circ$$

$$= 10 \times 0{\cdot}5 - 20 \times 0{\cdot}866 - 15 \times 0{\cdot}707 + 5 \times 0{\cdot}5$$
$$= -20{\cdot}42 \text{ N.}$$

∴ Sum of horizontal components is equal to 20·42 N to the left.

Resolve vertically: Vertical component $= P \sin \theta$

Sum of components

$$= 10 \sin 60° + 20 \sin 150° + 15 \sin 225° + 5 \sin 300°$$
$$= 10 \sin 60° + 20 \sin 30° - 15 \sin 45° - 5 \sin 60°$$
$$= 10 \times 0{\cdot}866 + 20 \times 0{\cdot}5 - 15 \times 0{\cdot}707 - 5 \times 0{\cdot}866$$
$$= +3{\cdot}73 \text{ N.}$$

∴ Sum of vertical components is equal to 3·73 N upward. These two components can be added together vectorially to give the resultant force.

$$\text{Magnitude of resultant} = \sqrt{3{\cdot}73^2 + 20{\cdot}42^2}$$
$$= 20{\cdot}75 \text{ N}$$
$$\tan \theta = \frac{3{\cdot}73}{20{\cdot}42}$$
$$\theta = 10° \, 21'$$

or, since the resultant is obviously in the 2nd quadrant

$$\theta = 169° \, 39'$$

∴ **resultant is 20·75 N at 169° 39′ to horizontal.**

Example 13.3

A uniform ladder, 10 m long, weighing 200 N, rests against a smooth vertical wall and upon a rough horizontal ground at a point 2·8 m from the foot of the wall. Determine the reaction of the wall and of the ground.

Forces acting on ladder:

(i) weight 200 N vertically down through mid-point of uniform ladder,
(ii) reaction of wall, R N, at right angles to wall, since this is smooth,
(iii) reaction of ground, P N, inclined since ground is rough.

METHOD 1

(1) All three forces must pass through one point. This point, Y, is determined by the intersection of the line of action of two of the forces, viz. (i) and (ii).

(2) Line of action of P must also pass through Y. Draw line through base of ladder and Y to obtain this direction. We now know the direction of the three forces and the magnitude of one of them, 200 N.

(3) Using Bow's notation, letter the spaces between the forces. The weight of the ladder will be represented by *ab*, the wall reaction by *ca*, and the ground reaction by *bc*.

(4) Draw, to suitable scale, the force diagram beginning with the weight of 200 N down. The intersection of the lines *bc* and *ac* gives the point *c*. *ca* is drawn parallel to R, and *bc* is drawn parallel to P.

(5) The length of *bc* represents the magnitude of the ground reaction P, and the length of *ac* represents the magnitude of the wall reaction R.

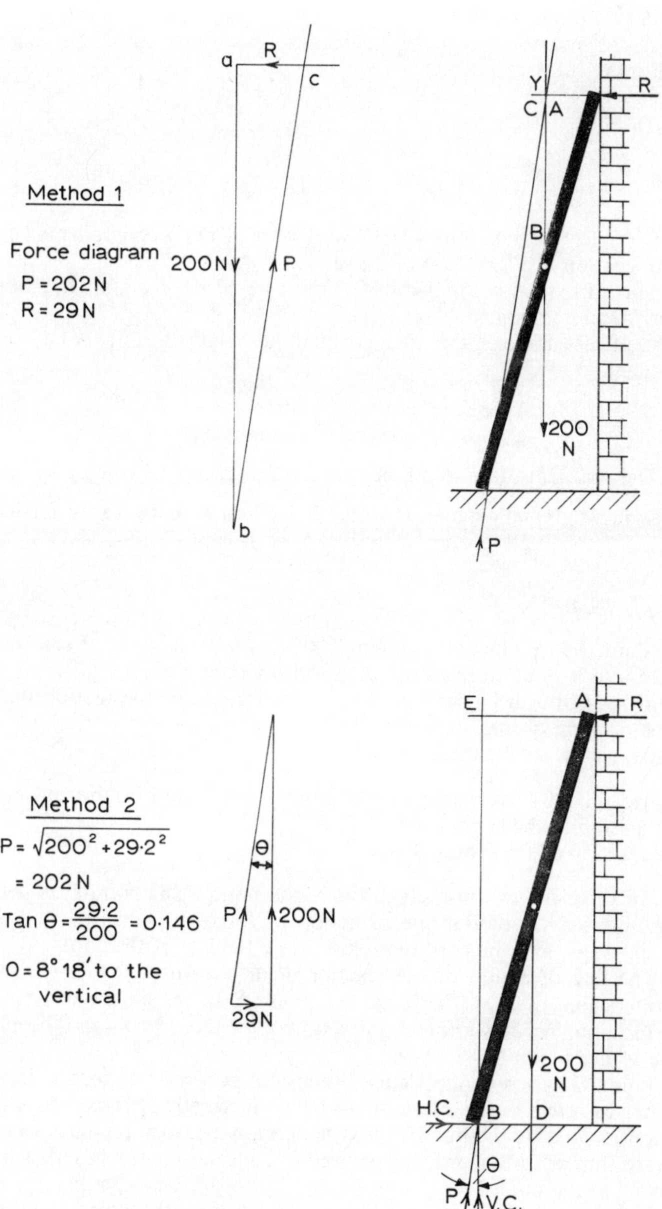

a
R
c
Method 1
Force diagram
P = 202 N
R = 29 N
200 N
P
b
Y
R
C A
B
200
N
P
Method 2
P = √200² + 29·2²
= 202 N
Tan θ = 29·2/200 = 0·146
θ = 8° 18′ to the
vertical
θ
P
200 N
29 N
E
A
R
200
N
H.C.
B
D
θ
P
V.C.

Figure 13.3

METHOD 2

(1) Take moments about B.

Clockwise moments equal anticlockwise moments, since the ladder is in equilibrium.

$$R \times BE = 200 \times BD$$

$$R = \frac{200 \times 1{\cdot}4}{9{\cdot}6}$$

$$= 29{\cdot}2 \text{ N.}$$

(2) Consider the horizontal (HC) and vertical (VC) components of the unknown reaction P at the base of the ladder.

(3) The horizontal component of P must equal the reaction R, since W has no horizontal component.

The vertical component of P must equal the weight W, since R has no vertical component.

Therefore, horizontal component = 29·2 N
vertical component = 200 N.

(4) The resultant of these can be found either graphically or by calculation.

Resultant ground reaction P = 202 N inclined at 8° 18′ to the vertical.
Wall reaction R = 29·2 N horizontal.

Example 13.4

A trapdoor, 3·5 m long and weighing 500 N, is fastened to a vertical wall with a hinge, and is supported at its other end with a cord 4 m long fastened to a support in the wall 5 m higher than the hinge. Determine the tension in the cord and the reaction of the hinge. (Fig. 13.4.)

Forces acting on the trapdoor are:

(i) weight 500 N vertically down through mid-point of uniform trapdoor,
(ii) tension in the supporting cord, T N,
(iii) reaction of the hinge, R N.

(1) All three forces must act through one point. This point E is determined by the intersection of the line of action of forces (i) and (ii). The weight is vertically down and the cord tension is along the line of the cord.

(2) The line of action of the reaction R must also pass through this point E. Draw a line through the hinge and point E to obtain the direction of the hinge reaction. We now know the direction of the three forces and the magnitude of one of them, 500 N.

(3) Using Bow's notation, letter the spaces between the forces. Notice that the letter M refers to the whole space between the 500 N force and the tension T as represented by the line EC, and not simply between the lines ED and EB. In a case similar to this one, it is as well to consider the forces either beginning or ending at the point E.

The weight of the trapdoor is represented by ms, the hinge reaction by sr, and the cord tension by rm.

(4) Draw, to a suitable scale, the force diagram beginning with the weight of 500 N down. The intersection of the lines sr and mr gives the point r. sr is

drawn parallel to the direction of the hinge reaction. *rm* is drawn parallel to the cord.

(5) The length of *sr* represents the magnitude of the hinge reaction and the length of *rm* represents the magnitude of the cord tension.

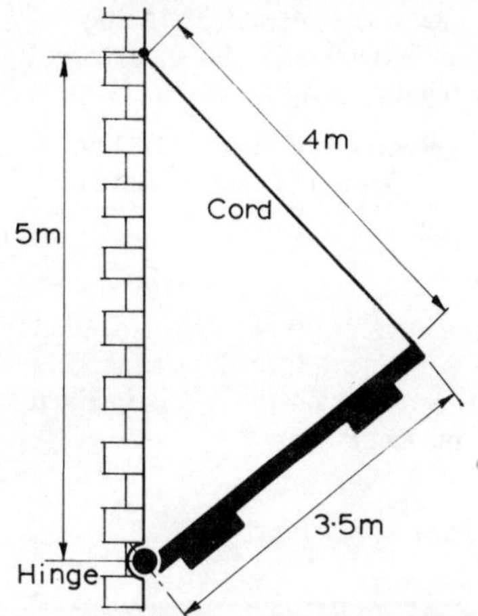

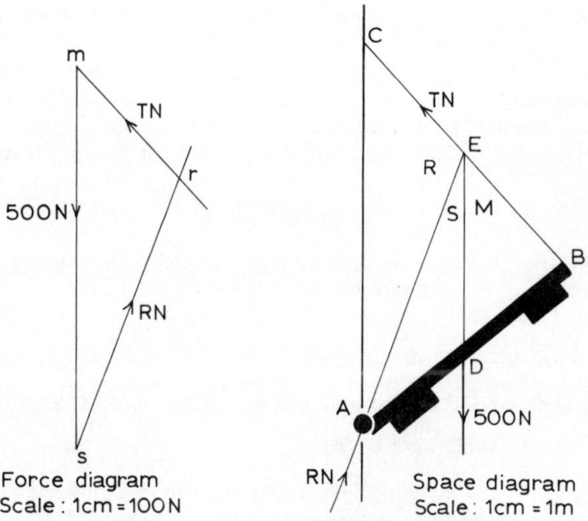

Figure 13.4

Solution by moments

The tension in the cord can be obtained by taking moments about the hinge A. The perpendicular distances of the 500 N force and the cord tension from the hinge would have to be measured.

The hinge reaction could be calculated by taking moments about the end B of the trapdoor. Again the perpendicular distances from B to the line of action of the force and the hinge reaction would have to be measured from the drawing.

Reaction of hinge = 382 N.
Tension in cord = 200 N.

Example 13.5

A uniform rod 1·7 m long weights 20 N. It is suspended from a fixed point by a string fastened to its ends, the length of the strings being 0·8 and 1·5 m respectively. Determine the angle at which the rod is inclined to the vertical and the tension in the strings (Fig. 13.5).

METHOD

(1) Forces acting on the rod:
(i) Pull P_1 exerted by short cord.
(ii) Pull P_2 exerted by long cord.
(iii) Weight 20 N vertically down through mid-point of uniform rod.

(2) These three forces must pass through one point. This point is determined by the intersection of the line of action of forces (i) and (ii), which is the fixed point of support D.

(3) The line of action of the weight of the rod must also pass through this fixed support D.

But the line of action of the weight passes through the mid-point C of the uniform rod and is also a vertical line.

(4) Hence the rod will tilt to such an angle which will make CD vertical.

(5) Measure angle CDB. Redraw the diagram commencing with a vertical line CD, and then make angle CDB equal to that obtained in the first diagram. Complete the triangle ABD which will give the correct position taken up by the rod.

(6) Measure the angle which AB makes with the vertical. The answer, by measurement, is 56°.

(7) The tension in the cords can be found either by drawing a force triangle for the three forces or by the principle of moments.

Solution by moments

Tension P_1 in string DB can be obtained by taking moments about A.

NOTE. P_2 has no moment about A.

Also, angle ADB is a right angle; hence the perpendicular distance from A to the line of action of force is given by the length AD = 1·5 m. If angle ADB were not 90°, the perpendicular distance would have to be measured.

By measurement AE = 0·71 m = BF

$$20 \times AE = P_1 \times 1{\cdot}5$$

$$P_1 = \frac{20 \times 0{\cdot}71}{1{\cdot}5} = 9{\cdot}4 \text{ N}$$

Similarly, moments about B will give the force P_2:

$$20 \times BF = P_2 \times BD$$

$$20 \times 0{\cdot}71 = P_2 \times 0{\cdot}8$$

$$P_2 = \frac{20 \times 0{\cdot}71}{0{\cdot}8} = 17{\cdot}6 \text{ N.}$$

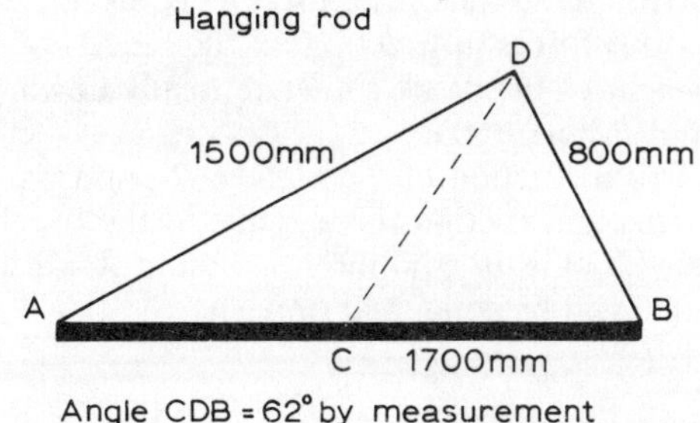

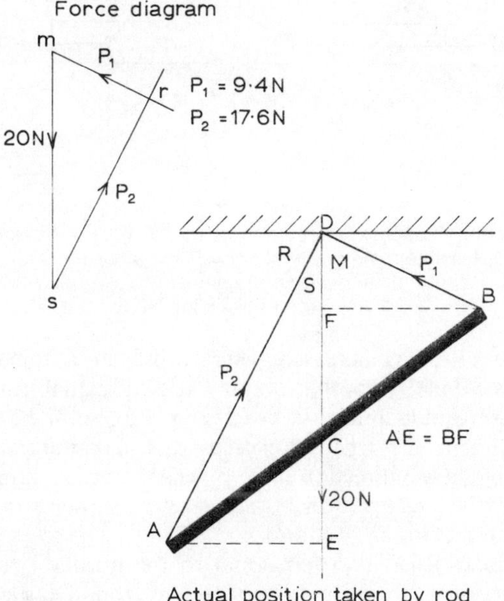

Figure 13.5

Graphical solution

The triangle of forces for the three forces acting on the rod is shown as *msr*. Commencing with the vertical weight of 20 N, the sides of the triangle are drawn parallel to the corresponding forces. The length of *mr* and *sr* to scale will represent the magnitude of the forces P_1 and P_2.

The rod is inclined at 56° to the vertical.
The tension in the 0·8 m string is 9·4 N.
The tension in the 1·5 m string is 17·6 N.

Friction

A horizontal force *P* N is applied to a block of weight *W* N resting on a rough horizontal surface. The force *P* is insufficient to move the block. There are four forces now acting on the block:

(1) The weight of the block acting vertically down.
(2) The pull *P* horizontal.
(3) The vertical reaction of the surface *R* upwards.
(4) The horizontal friction force *F* due to the roughness of the surface.

Since the block is in equilibrium, *W* and *R* will be equal and opposite, and *F* and *P* will be equal and opposite.

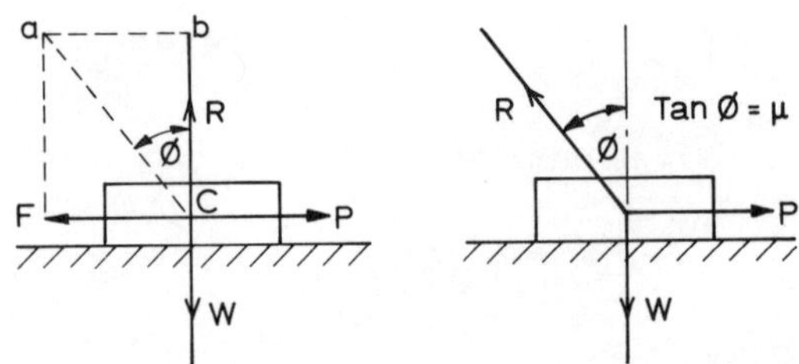

Figure 13.6 *Friction*

The *force* of friction *F* opposes the motion of the block. *F* increases as *P* increases, until *F* reaches a maximum value beyond which it cannot go. This maximum value depends upon the nature of the surfaces in contact, the two materials in contact and the weight of the sliding block.

Now let the pull *P* be increased. There will be an automatic and corresponding increase in *F* such that *F* and *P* are still equal. But whilst *P* can go on increasing indefinitely, *F* reaches a maximum beyond which it cannot go. And in fact when *P* exceeds this maximum value of *F*, the block will move in the direction of *P*. So that in order to move the block with a uniform velocity, it will be necessary to apply a force equal and opposite to the maximum friction force *F*.

This friction force *F* is proportional to the normal reaction between the two surfaces (which in the case of the horizontal surfaces is equal to the weight of the block) and is dependent upon the nature of the surfaces in contact.

$$\frac{F}{W} = \mu, \text{ the coefficient of friction of the surfaces in contact.}$$

$$\therefore \quad F = \mu W$$

or more general, $F = \mu \times$ normal reaction.

Now refer again to Fig. 13.6. The forces R and F are both forces which the surface exerts on the block and can therefore be combined together to form a single reaction. This reaction will no longer be normal to the surface, *but inclined backwards* (*relative to the direction of motion*) *through angle* ϕ.

$$\tan \phi = \frac{ab}{bc} = \frac{F}{R} = \frac{F}{W} = \mu$$

For a rough surface, the reaction is inclined backwards through an angle ϕ such that $\tan \phi = \mu$, the coefficient of friction. The angle ϕ is known as the **angle of friction.**

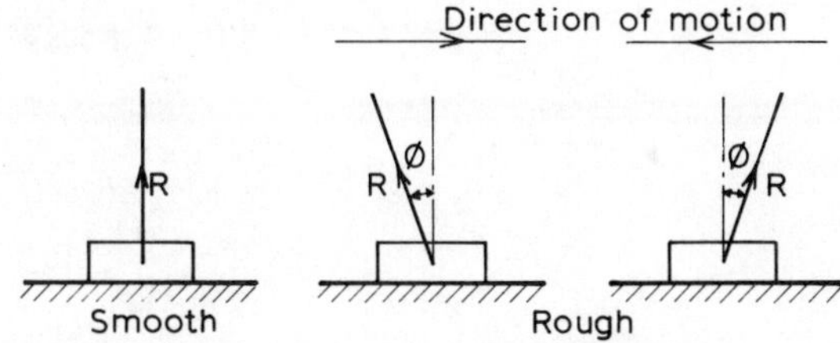

Figure 13.7 *Smooth and Rough Surfaces*

For a smooth surface the reaction of the surface is always at right angles to the surface. For a rough surface the reaction is inclined backwards to the direction of motion. The tangent of the angle of inclination is given by the coefficient of friction of the materials in contact.

Average Values of Coefficient of Friction

Materials	Coefficient of friction
bronze on bronze, dry	0·20
bronze on bronze, lubricated	0·05
leather on steel	0·55
wood on steel	0·50
steel on steel, lubricated	0·10

Friction and the inclined plane

Motion up the plane. Fig. 13.8

To determine the value of a force P needed to pull a body of weight W up a rough plane inclined at angle θ to the horizontal:

(*a*) when the pull is parallel to the plane, and

(*b*) when the pull is horizontal.

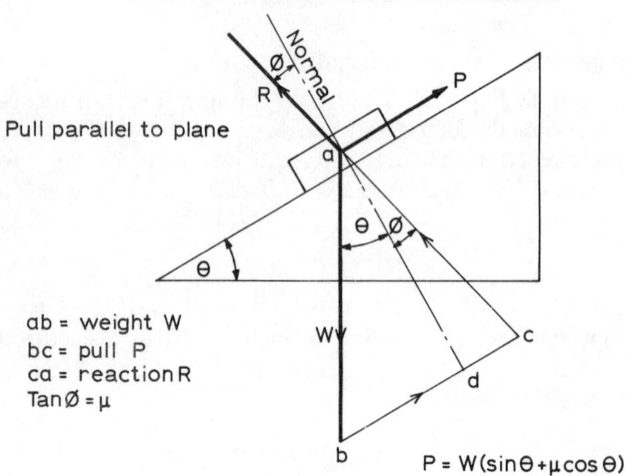
Motion up inclined plane with friction
Normal
Ø
R
P
Pull parallel to plane
a
θ
Ø
θ
ab = weight W
bc = pull P
ca = reaction R
TanØ = μ
W
c
d
b
P = W(sinθ + μcosθ)

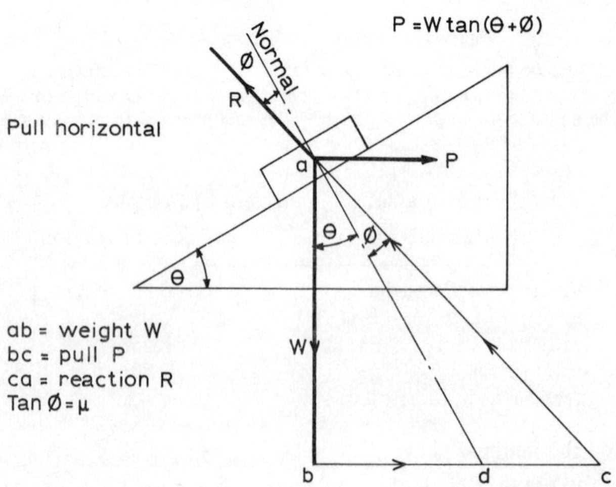
P = W tan(θ + Ø)
Normal
Ø
R
Pull horizontal
a
P
θ
Ø
θ
ab = weight W
bc = pull P
ca = reaction R
Tan Ø = μ
W
b
d
c

Figure 13.8

(*a*) *Pull parallel to the plane*

Three forces are acting on the body to produce equilibrium:

(1) pull P,
(2) weight W,
(3) reaction of plane inclined backwards, relative to the plane through angle ϕ.

abc is the triangle of forces in which:

ab represents to scale W N drawn vertically down.
bc represents to scale P N drawn parallel to the plane.
ca represents to scale R N drawn at angle ϕ to the normal.

The magnitude of P can be obtained either by calculation or by scaling the line *bc* on the diagram.

By calculation

$$\begin{aligned} bc &= bd + dc \\ &= W \sin\theta + ad \tan\phi \\ &= W \sin\theta + W \cos\theta \tan\phi \\ &= W(\sin\theta + \cos\theta \tan\phi) \\ P &= W(\sin\theta + \mu\cos\theta) \text{ since } \tan\phi = \mu \end{aligned}$$

Note with no friction, i.e. $\mu = 0$

$$P = W \sin\theta.$$

(*b*) *Pull horizontal*

The same three forces are acting as above. The directions of W and R are as above, but the direction of P is now horizontal.

abc is the triangle of forces in which:

ab represents to scale W N drawn vertically down,
bc represents to scale P N drawn horizontally,
ca represents to scale R N drawn at angle ϕ to the normal.

The magnitude of P can be obtained either by calculation or by scaling the line *bc* on the diagram.

$$\begin{aligned} bc &= ab \tan(\theta + \phi) \\ P &= W \tan(\theta + \phi) \end{aligned}$$

Note with no friction, i.e. $\mu = 0$ and also $\phi = 0$

$$P = W \tan\theta.$$

Motion down the plane (Fig. 13.9)

In considering the motion of the body down the plane, we shall see that there are two distinct problems which arise:

(1) The angle of the plane may be so steep that the body has to be held on the plane by a force P opposing motion.
(2) The angle may be so small that the body has to be pushed down the plane by a force P acting in the direction of the motion.

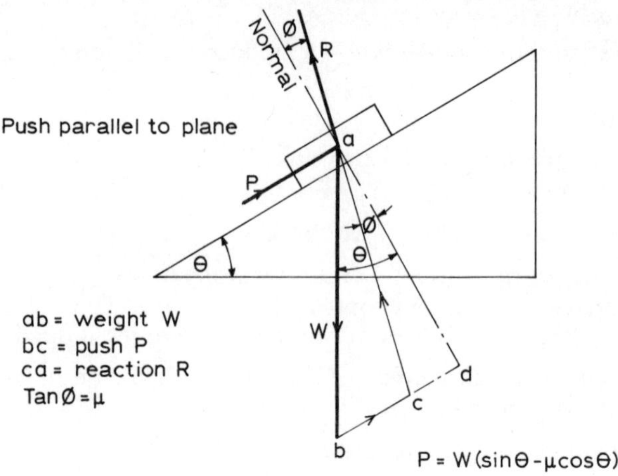

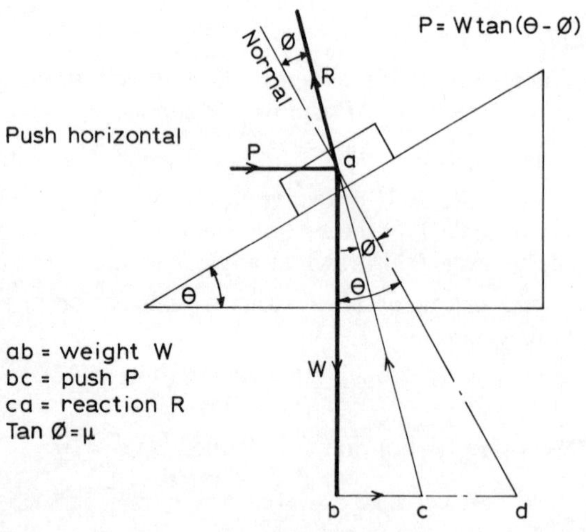

Figure 13.9

We shall consider the first case in detail, and then determine how the results may be applied to the second case.

To determine the value of a force P required to hold a body of weight W on a rough plane inclined at angle ϕ to the horizontal:

(*a*) when the force is parallel to the plane,
(*b*) when the force is horizontal.

(*a*) *Force parallel to the plane*

Three forces are acting on the body to produce equilibrium:

(1) push P,
(2) weight W,
(3) reaction of plane inclined backwards, relative to the plane through angle ϕ. Notice the difference between the direction of R when moving up the plane and that in the previous example.

abc is the triangle of forces in which:

ab represents to scale W N drawn vertically down,
bc represents to scale P N drawn parallel to the plane,
ca represents to scale R N drawn at angle ϕ to the normal.

The magnitude of P can be obtained either by calculation or by scaling the line *bc* on the diagram.

By calculation:

$$\begin{aligned} bc &= bd - cd \\ P &= W \sin \theta - ad \tan \phi \\ &= W \sin \theta - W \cos \theta \tan \phi \\ &= W(\sin \theta - \cos \theta \tan \phi) \\ P &= W(\sin \theta - \mu \cos \theta) \end{aligned}$$

(*b*) *Force horizontal*

The same three forces are acting as above. The directions of W and R are as above, but the direction of P is now horizontal.

abc is the triangle of forces in which:

ab represents to scale W N drawn vertically down,
bc represents to scale P N drawn horizontally,
ca represents to scale R N drawn at angle ϕ to the normal.

The magnitude of P can be obtained either by calculation or by scaling the line *bc* on the diagram.

$$\begin{aligned} bc &= ab \tan (\theta - \phi) \\ P &= W \tan (\theta - \phi) \end{aligned}$$

Angle of repose

The expression for the horizontal force P necessary to hold the block W on the plane is given by:

$$P = W(\sin \theta - \mu \cos \theta).$$

Taking $\cos \theta$ outside the bracket, and remembering that $\tan \theta = \dfrac{\sin \theta}{\cos \theta}$

we have:

$$P = W\cos\theta\,(\tan\theta - \mu)$$
$$= W\cos\theta\,(\tan\theta - \tan\phi)$$

Now if θ is greater than ϕ, then $(\tan\theta - \tan\phi)$ is positive and P is positive. This means that the force P must act up the plane. The plane is steep and the block would slide down the plane freely.

If θ is less than ϕ, then $(\tan\theta - \tan\phi)$ is negative and P is negative. This means that the slope of the plane is only small. The block will not move down without a force being applied to push it down the plane.

If θ is equal to ϕ then $(\tan\theta - \tan\phi)$ is zero and P is likewise zero. In this case the block is just on the point of slipping down the plane. The angle of the plane is such that a further increase in its value will cause the block to slip. This critical angle of the plane is known as the angle of repose, and its value is given by:

$$\tan\theta = \tan\phi = \mu\text{, the coefficient of friction.}$$

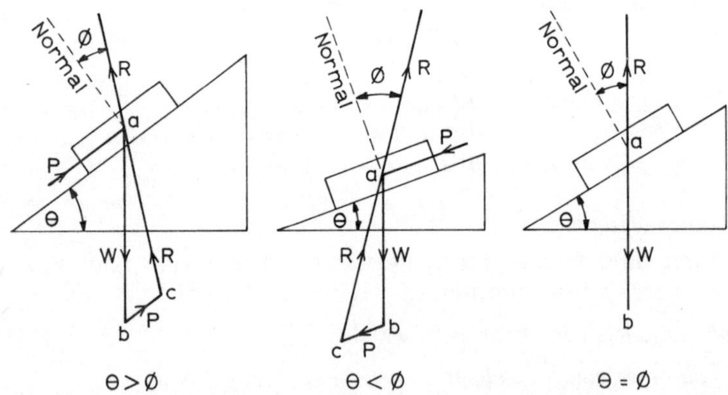

Figure 13.10 *Equilibrium on an Inclined Plane*

The question as to whether a block will remain at rest on an inclined plane depends upon the inclination of the plane and the angle of friction. If the angle of the plane is greater than the angle of friction, the block will immediately slide down the plane. If the angle of the plane is less than the angle of friction, the block will remain at rest on the plane and would in fact have to be pushed down the plane. The critical case is when the angle of the plane is equal to the angle of friction. This angle is sometimes referred to as the angle of repose.

If the angle of the plane is greater than the angle of repose, then the block will slide down the plane freely. This fact will have an important bearing upon problems concerned with screw threads, particularly applied to screw jacks and worm-gear drives.

Incidentally, there is a method here for determining experimentally the value of the coefficient of friction. Increase the angle of the plane until the block just slides down. Note the value of this angle. Then, using a larger angle of plane, gradually reduce this until the block no longer slides down freely. The average of the two angles will give the angle of repose, and the tangent of this angle will give the coefficient of friction.

Example 13.6

The force required to drag a body weighing 150 N at a constant speed along a horizontal table is 40 N. If the force is applied at 20° to the table, determine the coefficient of friction between the two surfaces.

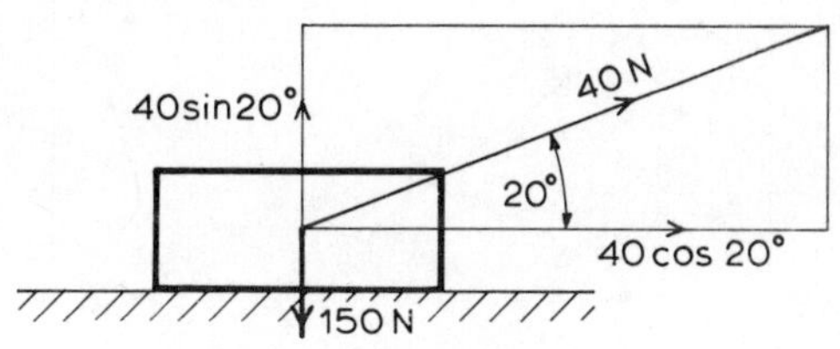

Figure 13.11

$$\begin{aligned}\text{The horizontal component of the 40 N force} &= 40 \cos 20^\circ \\ &= 40 \times 0{\cdot}9397 \\ &= 37{\cdot}6 \text{ N} \\ \text{Vertical component of the 40 N force} &= 40 \sin 20^\circ \\ &= 40 \times 0{\cdot}3420 \\ &= 13{\cdot}6 \text{ N} \\ \text{Normal reaction between block and table} &= 150 - 13{\cdot}6 \\ &= 136{\cdot}4 \text{ N.}\end{aligned}$$

In effect, we have a horizontal force of 37·6 N moving a load of 136·4 N.

$$\text{Coefficient of friction} = \frac{37{\cdot}6}{136{\cdot}4} = 0{\cdot}27.$$

Example 13.7

A horizontal force of 30 N is required to move a block weighing 80 N over a horizontal surface. What force would be required:
(*a*) to pull the block up an inclined plane,
(*b*) to push the block down an inclined plane,
if the inclination of the plane is 15° and the force acts parallel to the plane?

$$\begin{aligned}\text{The coefficient of friction between surfaces} &= 30/80 = 0{\cdot}375 \\ \text{The angle of friction } \theta \text{ is given by } \tan\theta &= 0{\cdot}375 \\ \theta &= 20^\circ\ 33'.\end{aligned}$$

Graphical solution (Fig. 13.12)

Draw the three forces, P, W and R, in position, with R inclined backward through angle 20° 33′ to the normal. This angle can either be measured or constructed so that $\tan\theta = \frac{3}{8}$. Measure 8 units along the normal ax and 3 units perpendicular to the normal xy. Join ay, which gives the direction of the reaction R.

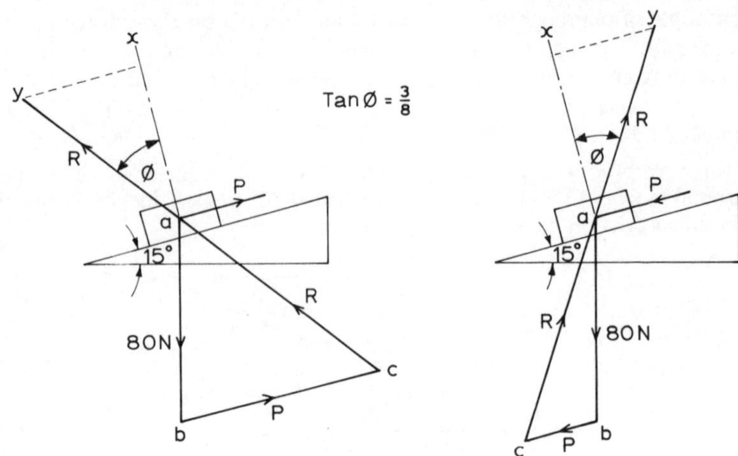

Figure 13.12

abc is the triangle of forces.
ab represents 80 N to scale.
bc is drawn parallel to *P*.
ac is drawn as a continuation of the reaction *R*.
bc measures 49·6 units, representing 49·6 N pull.
ca measures 82·4 units, representing 82·4 N reaction.

For motion down the plane: *R* is drawn on the other side of the normal, at 20° 33′, using the same construction as before.

abc is the triangle of forces.
ab represents 80 N to scale.
bc is drawn parallel to *P*.
ac is drawn as a continuation of the reaction *R*.
bc measures 8·3 units, representing 8·3 N pull.
ca measures 82·4 units, representing 82·4 N reaction.

Note that the direction of *bc* (and therefore of *P*) is opposite in the two cases.

By calculation

$$\text{Up the plane } P = W(\sin\theta + \mu\cos\theta)$$
$$= 80(\sin 15^\circ + \tfrac{3}{8}\cos 15^\circ)$$
$$= 80(0{\cdot}2588 + \tfrac{3}{8}\times 0{\cdot}966)$$
$$P = 49{\cdot}6 \text{ N.}$$

$$\text{Down the plane } P = W(\sin\theta - \mu\cos\theta)$$
$$= 80(\sin 15^\circ - \tfrac{3}{8}\cos 15^\circ)$$
$$= 80(0{\cdot}2588 - \tfrac{3}{8}\times 0{\cdot}966)$$
$$P = -\,8{\cdot}32 \text{ N.}$$

The negative sign means that the direction of *P* is down the plane, *i.e. the block has to be pushed down.*

Force to move up plane = 49·6 N
Force to move down plane = 8·32 N.

Example 13.8

A block weighing 20 N is placed on an inclined plane whose angle of inclination is increased until the block just slips down the plane.

This occurs at an angle of 28°. Determine the horizontal force to move the block:

(*a*) over a horizontal plane,
(*b*) up a plane inclined at 45° to the horizontal, and
(*c*) down a plane inclined at 15° to the horizontal.

Since the block just slips down the plane inclined at 28°, this angle is the angle of friction θ.

$$\mu = \tan\theta = \tan 28^\circ = 0{\cdot}53.$$

$$\text{Force to move block over horizontal plane} = \mu W = 0{\cdot}53 \times 20 = 10{\cdot}6\text{ N.}$$

Horizontal force to move block up plane inclined at 45°

$$\begin{aligned} P &= W \tan(\theta + \theta) \\ &= 20 \tan(45^\circ + 28^\circ) \\ &= 20 \tan 73^\circ \\ &= 20 \times 3{\cdot}27 \\ P &= 65{\cdot}4\text{ N.} \end{aligned}$$

Horizontal force to move block down plane inclined at 15°

$$\begin{aligned} P &= W \tan(\theta - \theta) \\ &= 20 \tan(15^\circ - 28^\circ) \\ &= 20 \tan(-13^\circ) \\ &= 20 \times -0{\cdot}23 \\ &= -4{\cdot}6\text{ N} \end{aligned}$$

the negative sign indicating that *block has to be pushed down the plane.*

Force to move over horizontal plane = 10·6 N
Force to move up 45° plane = 65·4 N
Force to move down 15° plane = 4·6 N.

Funicular polygon

Consider, now, three forces M, P and Q, which are all acting in one plane but which do not pass through the same point. We are asked to determine the magnitude, direction and position of the resultant of these forces. We can determine the magnitude and direction of the resultant by the usual method of drawing the force diagram. But before we can decide the position in which the resultant acts, it will be necessary to draw another diagram which is known as the **link** or **funicular** polygon.

If all the forces passed through one point, then of course the resultant would also pass through this point. In which case there would be no need to construct a link polygon.

The construction is carried out as follows (Fig. 13.13(i)):

(1) Draw the force diagram, *abcd*, in the usual way, and obtain the magnitude and direction of the resultant force *R* as given by the line *ad*.

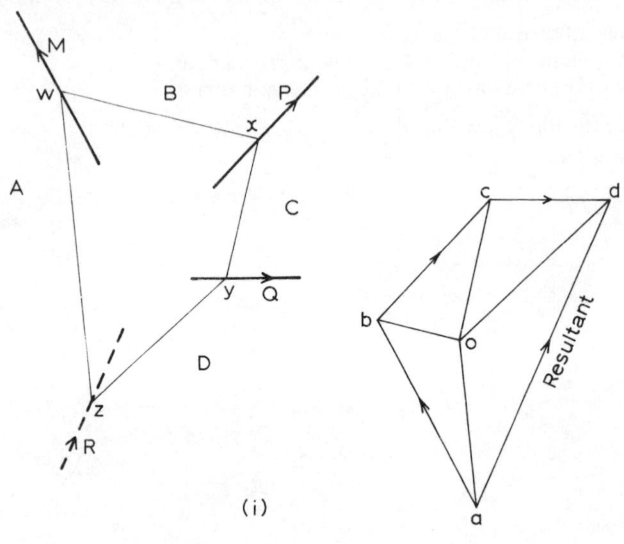

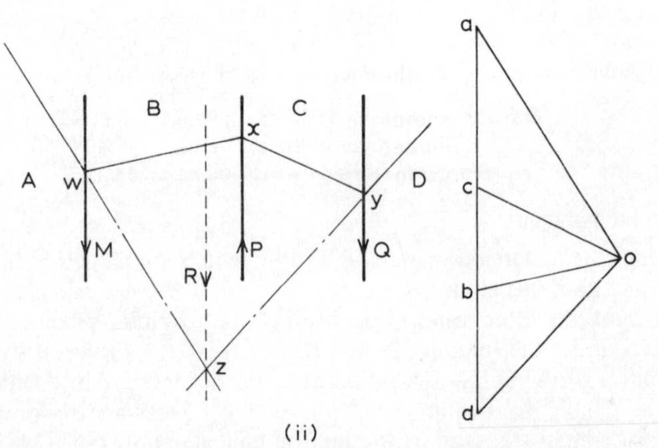

Figure 13.13

(2) Choose any point *o* either inside or outside the force diagram, and join *o* to each corner of the force diagram.
(3) From any point *w* in the line of action of force *M*, draw a line in space A parallel to the line *oa*.
(4) From the point *w* draw the line *wx* in space B parallel to line *ob* and cutting force *P* in the point *x*. (You may have to produce the line of action of the forces to get the necessary intersection.) This is continued until lines have been drawn in all the spaces. The lines parallel to *oa* and *od* can now be produced until they meet in *z*.
(5) The resultant of the forces will pass through the point *z*. Through *z* draw a line parallel to *ad* to represent the resultant force *R*.

The diagram *wxyz* is the **link polygon** or the **funicular polygon.**

Resultant of parallel forces (Fig. 13.13(ii))

The construction is similar to that outlined above. The point *o* must of course be outside the force diagram, which in this case is a vertical straight line. In order to get the intersection point *z*, it will normally be necessary to project the line in spaces *A* and *D* backwards.

For the example given, the resultant will have the magnitude and direction as given by the line *ad*, and its position will be given by having to pass through the point *z*.

Graphical determinations of beam reaction

The magnitude of the reaction of beam supports can be obtained graphically by an application of the funicular polygon.

Let us take the case of a beam loaded as shown in Fig. 13.14. The loads are *M*, *P* and *Q*, and the reactions *N* and *R*.

(1) Letter the spaces, using Bow's notation, and draw the force diagram (in this case a straight line *ad*).
(2) Choose any point *o* and draw the lines *oa*, *ob*, *oc* and *od*.
(3) From any point *w* on the line of action of *N* (produced for preference) draw a line *wx* parallel to *oa* in space A, cutting force *M* at *x*.
(4) From *x* draw the line *xy* in space B parallel to *ob*.
(5) Repeat this construction until the last line has been drawn, which in this case will intersect force *R* at *k*.
(6) Join the points *w* and *k*, and through *o* draw a line *oe* parallel to *wk*, cutting the vertical force line *ad* at *e*.
(7) Then *de* represents the magnitude and direction of reaction *R*, and *ea* represents the magnitude and direction of reaction *N*.

EXERCISE 13

1 The following forces act at a point:
(i) 16 N in direction due east,
(ii) 20 N in direction due north,

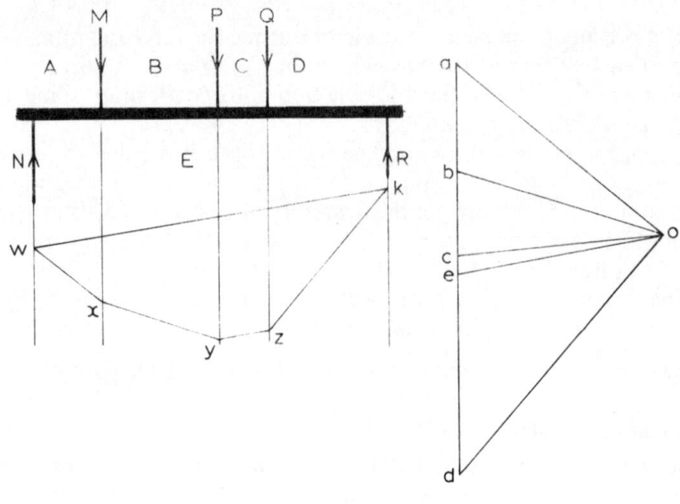

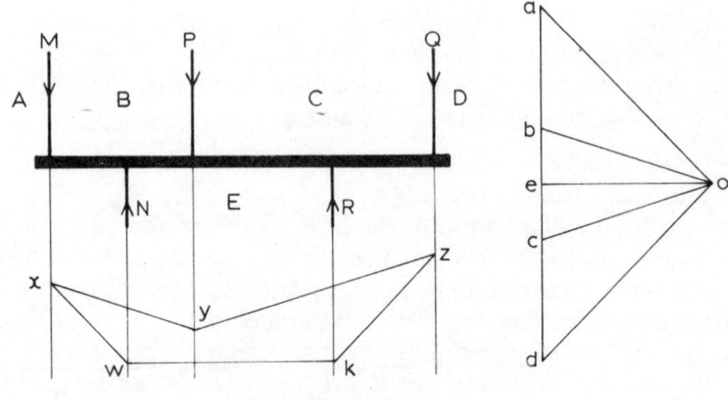

Figure 13.14

(iii) 30 N in direction north-west,
(iv) 12 N in direction 30° south of west.

Determine the magnitude and direction of the resultant of these forces:
(*a*) when they are all pulling away from the point, and
(*b*) when (i) and (iii) are acting towards the point and (ii) and (iv) are pulling away from the point.

2 AB is a uniform rod, 2·5 m long, weighing 200 N, which is hinged at B in a horizontal position by means of a cord attached at A and making 30° with

the horizontal. A load of 1000 N is carried at C, 2 m from B. What is the pull in the rope and the hinge reaction?

3 ABC is a wireless mast assumed to be rigid, which has a horizontal pull at the top, A, of 3000 N. The mast is kept vertical by a guy-wire attached to a point B, 10 m from the base C, this wire making an angle of 35° with the mast. If the mast is 18 m long, determine the pull in the guy-rope and the magnitude and direction of the ground reaction.

4 Determine the force required to pull a block of 200 N up a 30° plane when the pull is:

(*a*) parallel to the plane,
(*b*) horizontal.

The coefficient of friction can be taken as 0·2.

5 A uniform ladder 6 m long weighing 500 N is resting against a vertical wall. The ladder makes an angle of 30° with vertical. Determine the reaction of the wall which can be taken as smooth and also the magnitude and direction of the ground reaction.

6 A uniform rod AB 6 m long weighing 10 N is hinged to a wall at A, and is supported by a cord attached from B to the wall at C. The triangle ABC is equilateral. Determine the pull in the cord and the magnitude and direction of the reaction at the hinge.

7 A wall crane consists of a jib 6 m long and a tie-bar 4 m long. The points of attachment to the wall are 3 m vertically apart. If the crane supports a load of 40 kN and the rope runs parallel to the tie-bar, determine a suitable cross-section area of the tie-bar, given that the tensile stress is not to exceed 30 N/mm^2. [U.L.C.I.]

8 Three horizontal wires attached to the top of a pole have the following pulls and directions: 450 N due south, 500 N due east, 400 N north-east. Determine:

(*a*) the direction and magnitude of the resultant pull, and
(*b*) the cross-sectional area of the horizontal wire which would have to be fitted to the pole to balance the resultant pull, if the maximum stress in the wire is not to exceed 100 N/mm^2. [U.L.C.I.]

9 ABCD is a rectangle, AB being 6 m long and BC 8 m; E is a point in BC 4·5 m from B. Forces 3, 4, 5, 6 N act along AB, AE, AC, AD respectively. Find the magnitude and direction of their resultant by the application of either the parallelogram law or the triangle of forces. [N.C.T.E.C.]

10 A uniform bar 3 m long and weighing 4 N is suspended from a single point by cords of length 2 m and 2·5 m respectively attached to the ends of the bar. Determine the inclination of the bar to the horizontal when it is in equilibrium, and find the tension in each cord. [N.C.T.E.C.]

11 Forces equal to 3, 4, 6 and 7 N act in order round the 10 cm sides of a square ABCD. Calculate the magnitude of their resultant and its position measured from corner D.

12 Six forces act outwards along the lines joining the centres of a regular hexagon to the corners as shown in Fig. 13.15. The system is balanced by two forces, one acting in the direction of the force of 6 kN and the other in a direction perpendicular to this. Find the magnitude of the balancing forces. [N.C.T.E.C.]

13 State the general conditions of equilibrium of a body under the action of three co-planar forces.

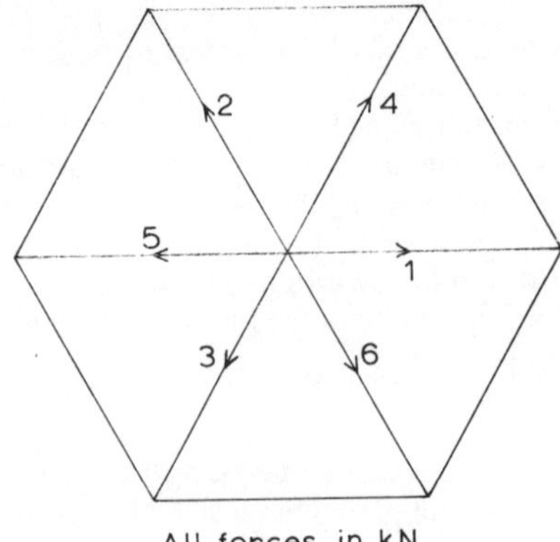

Figure 13.15

A uniform ladder 5 m long rests against a smooth vertical wall and upon a rough horizontal ground at a point 1 m from the foot of the wall. Make an accurate diagram showing the forces which maintain it in equilibrium; also state the values of the reactions of the wall and the ground in terms of the weight *W* of the ladder. [N.C.T.E.C.]

14 A uniform rod AB, of length 2 m and weight 8 N, is hinged at A to a vertical wall and the end B is connected by a horizontal cord to a point on the wall 0·8 m above A. Find the magnitude and direction of the force on the hinge and the tension in the cord. [N.C.T.E.C.]

15 Parallel forces of 1, 2, 3 and 4 N act at the corners A, B, C and D respectively of a square of 1 m side at right angles to the plane of the square. Find the magnitude of the resultant and the position of the centre of the forces (*a*) when they are like, (*b*) when the forces at A and C act in the opposite direction to those at B and D. [N.C.T.E.C.]

16 The force required to drag a body weighing 100 N at a constant speed up a plane inclined at 30° to the horizontal is 68 N, the force being parallel to the plane. Find, graphically or otherwise, the coefficient of friction and the angle of friction. [N.C.T.E.C.]

17 Define the term 'coefficient of friction' as applied to a body sliding along a plane surface.

A body weighing 50 N is dragged at a constant speed along a horizontal table by means of a rope inclined at 30° to the table. The tension in the rope is found to be 10 N. Determine the coefficient of friction between the two surfaces. [N.C.T.E.C.]

18 A case of weight 0·5 kN is pulled along a horizontal floor with uniform velocity. If the coefficient of friction between the case and the floor is 0·3, find the pull exerted when applied (*a*) horizontally, (*b*) at 30° to the horizontal. [N.C.T.E.C.]

19 In an experiment to determine the coefficient of friction between a brake fabric and mild steel, a block lined with material was drawn across a horizontal steel table. The following results were obtained:

Normal load between surfaces (N)	1	6	11	14	18	20
Horizontal forces required (N)	0·8	2·44	4·4	5·56	7·24	8

Plot these figures on a graph and hence determine the coefficient of friction. [N.C.T.E.C.]

20 A brake drum 1 m in diameter is rotating at 400 rev/min and the block is lined with material whose coefficient of friction is 0·4. Calculate the work done in bringing the drum to rest in two minutes if the normal force between block and drum is 200 N. [N.C.T.E.C.]

21 Define: (*a*) coefficient of friction, (*b*) angle of friction.

A brake drum 1 m in diameter, is fitted with a single block of brake material with a coefficient of friction 0·4. A maximum normal force of 200 N may be applied to the block and under a given set of conditions it is found that this is just sufficient to hold the drum at a steady speed of 50 rev/min. Determine: (*a*) the torque on the drum shaft, and (*b*) the power absorbed in friction. [N.C.T.E.C.]

22 State the factors affecting the friction between two plane dry surfaces when the speed of rubbing is low.

Describe briefly experiments which may be carried out to verify each of these factors.

A machine table of weight 1500 N moves horizontally on its bed. If the coefficient of friction between table and bed is 0·05, what force is require to move the table when a casting weighing 10 kN is carried on it? [N.C.T.E.C.]

23 Describe an experiment by which the coefficient of friction between two surfaces may be determined by allowing a slider to descend an inclined plane.

Illustrate your answer by sketches and derive the expression relating to the coefficient of friction.

Also state the effect of varying

(*a*) the weight on the slider, and

(*b*) the area of the face of the siider. [U.E.I.]

Chapter Fourteen
Frameworks

XYZ is a simple frame loaded at X as shown in Fig. 14.1. The frame rests on supports at Y and Z, which can be taken to provide vertical reactions. Since the frame is isosceles, the point X is midway between Y and Z, so that the reactions at these points are equal to half the load at X.

Now before the frame can be designed, it will be necessary to know the magnitude of the forces which are acting in each member of the frame. When this is known, it is a simple matter to calculate the cross-sectional area of the particular member, and therefore to determine the necessary size of section required to carry the load safely.

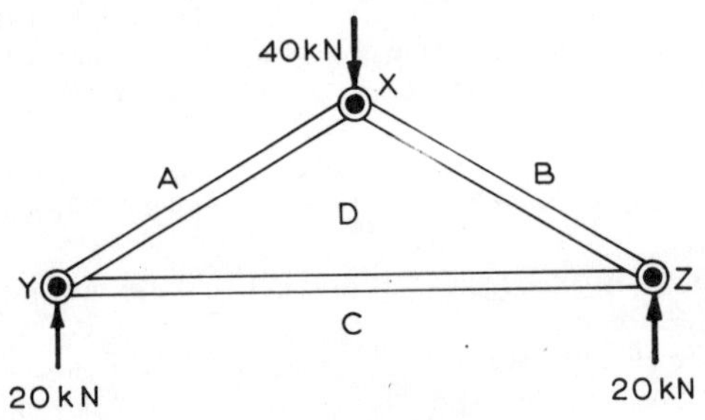

Figure 14.1

To help us in identifying the forces, we letter the spaces between the forces, using Bow's notation. The capital letters A, B, C and D are placed in the spaces between the forces, so that the force in member XY will be referred to as force *ad* and the reaction at Z as the force *bc*.

Consider the forces acting at X. They are three in number:

(1) the 40 kN load, *ab*,
(2) the force in YX, *da*,
(3) the force in ZX, *bd*.

We know the magnitude and direction of the first force and the direction of the other two forces. In fact, you have a very similar problem to the type you will have dealt with at an earlier stage.

Let us isolate the forces which are acting at X. They are shown in Fig. 14.2, which also shows the triangle of forces *abd* drawn in the usual manner.

By measurement: force *bd* is equal to 40 kN
force *da* is equal to 40 kN

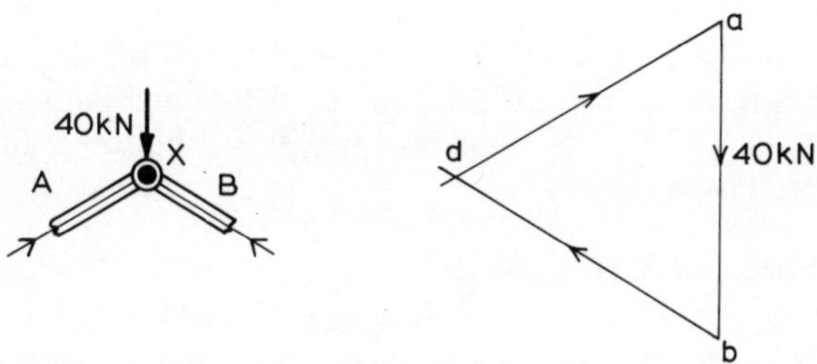

Figure 14.2

Similarly, we can deal with the forces acting at the point Y. These forces are:

(1) upward reaction 20 kN, *ca*,
(2) the force in XY, *ad*,
(3) the force in YZ, *dc*.

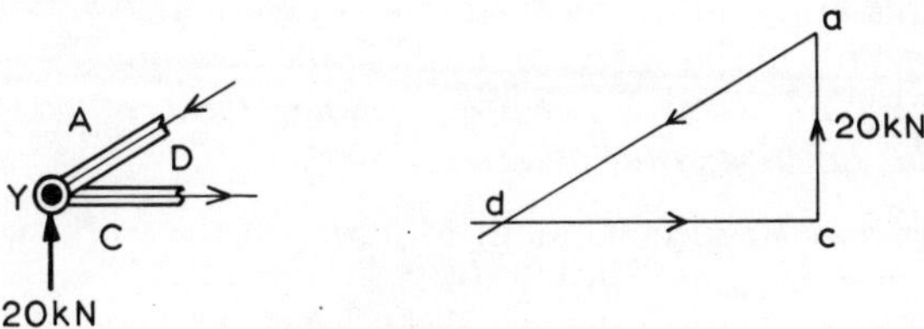

Figure 14.3

These forces are shown isolated in Fig. 14.3, and the triangle of forces *adc* drawn as shown.

By measurement: force *ad* is equal to 40 kN
force *dc* is equal to 34 kN

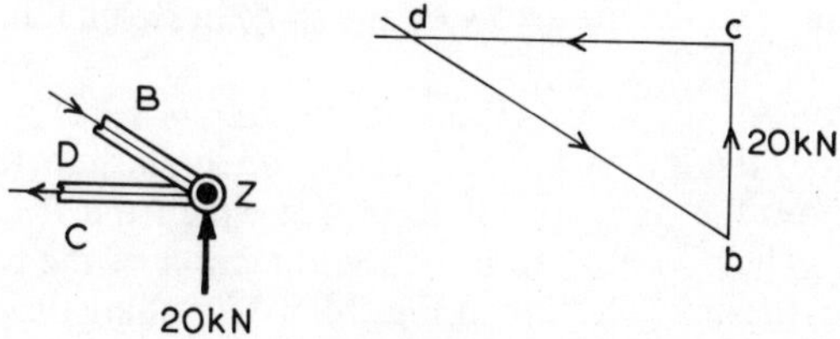

Figure 14.4

Finally, the forces acting at the point Z are shown isolated in Fig. 14.4, together with the force diagram *bcd*.

By measurement: force *db* is equal to 40 kN
force *cd* is equal to 34 kN.

Note on direction of forces

In the drawing showing the isolated forces, you will see that arrows have been inserted to show the directions in which the forces are acting. These directions have in all cases been obtained from the force diagram. The method is simply as follows:

Example: Forces at Y (Fig. 14.3)

Direction of one force is always known. In this case it is the direction of the reaction, upwards. Put the arrow in the force diagram on the line *ca* (which represents the 20 kN reaction). This arrow should point in the upward direction. Now, using this as the starting direction, put arrows on the remaining lines so that the *arrows all point in the same direction round the triangle*—in this case in the anti-clockwise direction. Then, having fixed the arrows on the force diagram, transfer them to the space diagram, making sure that the direction on the space diagram is the same as that on the force diagram. Thus the arrow on the horizontal link points to the right because the arrow in the force diagram points towards the right. Similarly, the arrow in the inclined link is pointing down and to the left, because it points in this direction in the force diagram.

This is very important. Remember, *we know the direction of one and can then find the direction of the others.*

There are two important things to notice in the work which we have just done in connection with Fig. 14.1.

First, we have obtained the magnitude of the force in each link—twice.

For example: from Fig. 14.2 we have force $da = 40$ kN
from Fig. 14.3 we have force $ad = 40$ kN.

Same force, same magnitude, but opposite directions.

Second, the arrows on the ends of the link. *The arrow at one end of any link points in the opposite direction to the arrow at the other end.*

For example: the arrow on the horizontal link at the point Y points towards the right, whilst the arrow on the same horizontal link at the point Z points towards the left.

The directions of the arrows indicate the type of force in the link; they indicate whether the link is in compression or in tension. Remember that the arrows show the direction of the force which the *link exerts on the junction point*. They do not represent the direction of the force which the point exerts on the link. The link in Fig. 14·5(i) is pulling the points towards each other. This means that the two points at the end of the link are trying to move away from each other; in other words, they are stretching the link. A link which is subjected to stretching is called a tie. The arrows on the link in Fig. 14·5(ii) show that the link is pushing the points apart. This means that the two points at the end of the link are trying to come together; in other words, they are compressing the link. A link which is subjected to compression is called a strut.

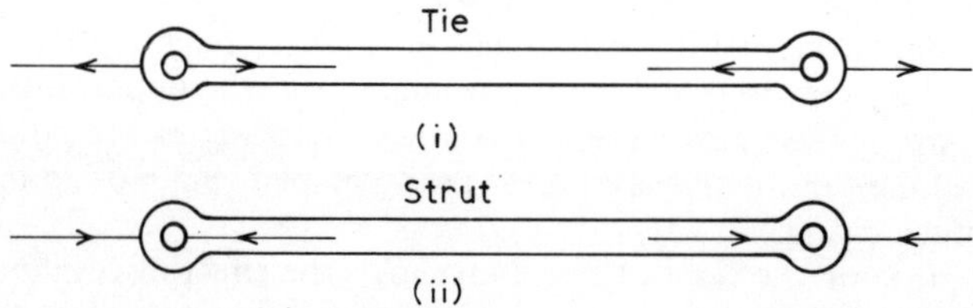

Figure 14.5 *The Difference between the Tie and the Strut*

The tie exerts an inward pull at either end to prevent the pins being pulled apart by the tensile loads.

The strut exerts an outward force at either end to prevent the compressive loads pushing the pins together.

A strut must always be a solid, rigid piece of material. A tie can either be rigid or it may be flexible like a rope.

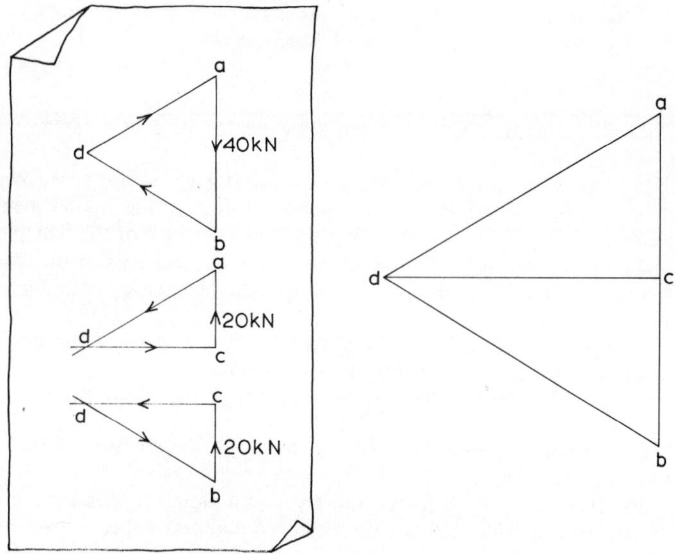

Figure 14.6 *Force Diagram for the Simple Roof Frame*

Now let us look at Fig. 14.6. The left-hand portion of the diagram shows a reproduction of the forces diagrams of Figs. 14.1–14.3. The lettering is the same as that used in these diagrams. Imagine the lower two triangles on the left moved so that the two horizontal lines came together so that the middle triangle rests on the bottom triangle. The length of *dc* is the same in each case. The angle at *c* is a right angle in each case. So the two points *c* would come together and the two points *d* would come together. And the line *acb* would be a straight vertical line whose length represents 40 kN to the scale of the drawing.

Now, keeping the two lower triangles in contact, imagine that they are moved up the paper to cover the top triangle. And the important thing is that they would just cover the top triangle. The angles are equal. The length *ab* of the top triangle is equal to the combined length of *ac* and *cb* of the two other triangles.

The diagram on the right of Fig. 14.6 shows the effect of combining the three triangles on the left. The only apparent difficulty is the direction of the arrows. So for the time being the arrows are omitted from the diagram on the right.

Therefore, we see that the problem of the forces in the frame could be solved without isolating the forces acting at each point in turn and also with the use of one force diagram only. The complete solution using the single force diagram is worked out below.

Example 14.1

Determine the magnitude and nature of the forces acting in the members of a simple frame loaded with a single load of 40 kN as shown in Fig. 14.7.

METHOD

(1) Draw to scale the space diagram showing the direction of the various members of the frame.

(2) Using Bow's notation, indicate with a capital letter the space between each set of forces, remembering that there are forces acting in the members, and therefore a letter, D in this case, is placed in the centre of the triangle.

(3) Calculate the reactions of the supports. In this case, since the frame is isosceles and the 40 kN load acts at the mid-point, the reactions are both equal to 20 kN.

(4) Choose a suitable scale for the force diagram, and draw the line *ab* to represent in magnitude and direction the 40 kN force.

(5) From the point *b* draw the line *bc* to represent the reaction of 20 kN on the right.

(6) From the point *c* draw the line *ca* to represent the reaction of the 20 kN support on the left.

This completes the external forces and gives us a closed force diagram. Since all the forces are vertical, in this case the force diagram appears as a straight line.

(7) Now deal with the force AD in the left-hand sloping member. Through *a* draw a line parallel to this member. The length of the line is not known because we do not yet know the magnitude of the force AD.

(8) Through *b* draw a line parallel to the force BD in the right-hand sloping member. Continue the line until it intersects the one drawn in 7.

(9) Through *c* draw a line parallel to the force CD in the horizontal member. This line should pass through the point *d*, which serves as a check on the accuracy of your drawing and the correctness of the method.

(10) Measure *ad*, *cd* and *bd* to scale:

$$ad = 40 \text{ kN},$$
$$bd = 40 \text{ kN},$$
$$c = 34{\cdot}6 \text{ kN}.$$

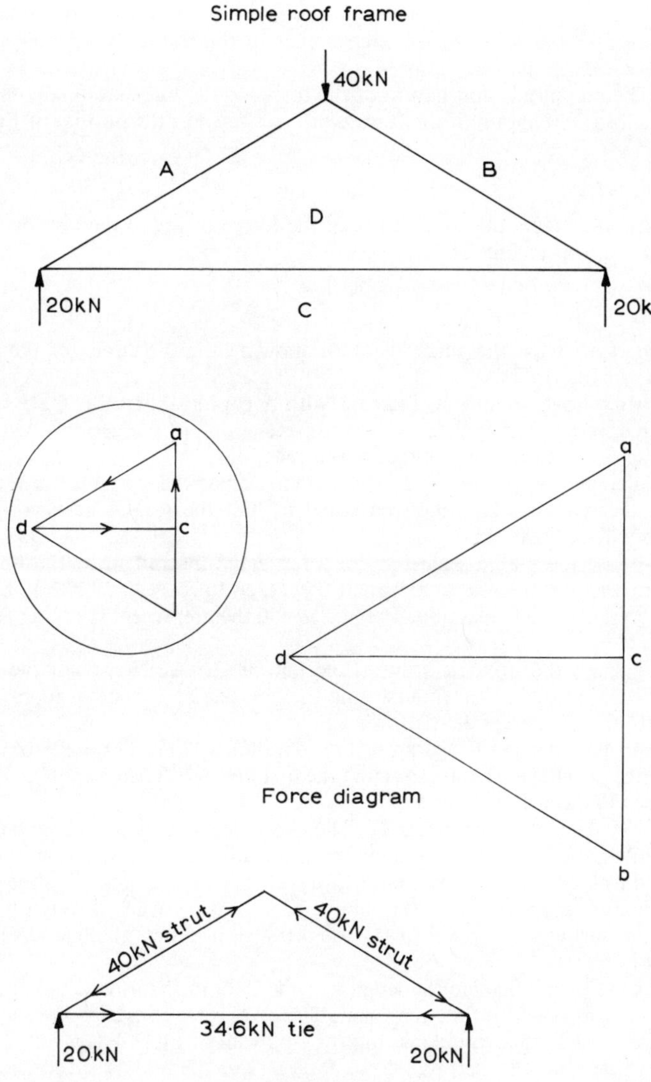

Figure 14.7

(11) Finally, we have to determine the nature of the forces. As an example, consider the forces acting at the left-hand support. The forces are *ca*, *ad* and *dc*. The direction of *ca* is known—upwards.

Beginning at *c*, trace out a path from *c* to *a*, from *a* to *d*, and then from *d* back again to *c*. The direction in which we have moved is indicated in the smaller force diagram drawn inside the circle. Now transfer these arrows on to the corresponding links on the space diagram as shown at the bottom of Fig. 14.7.

In this case, we have dealt only with the forces at the left-hand support. The procedure must be repeated for each corner of the frame. *Always begin in the direction as given by a known force.*

(12) The magnitude and nature of the forces in the links are finally indicated on a small-scale diagram of the framework as shown at the bottom of Fig. 14.7.

Example 14.2

Determine the magnitude and nature of the forces acting in the members of the roof truss shown in Fig. 14.8.

METHOD

(1) Draw to scale the space diagram showing the directions of the various members of the frame.

(2) Using Bow's notation, indicate with a capital letter the space between each force, remembering that there are forces acting in the various members of the frame besides the external forces and reactions.

(3) Calculate the reactions at each support. Since the frame is symmetrically loaded, the reactions will both be equal to half the total load carried. Each reaction will be 2000 N.

(4) Choose a suitable scale for the force diagram, and draw the line *ab* to represent 1600 N, then *bc* to represent 800 N, *cd* to represent 1600 N, and then *de* to represent 2000 N reaction. The line *ea* will then represent the other reaction, 2000 N.

(5) Through the point *a* draw a line parallel to AF. Through the point *e* draw a line, parallel to EF, intersecting the first line at *f*. Then *af* represents the force AF and *ef* represents the force EF.

(6) Through the point *f* draw a line parallel to FG. Through the point *b* draw a line parallel to BG, intersecting the first line at *g*. Then *bg* and *fg* represent the forces BG and FG.

(7) Next deal with the space H. The adjoining spaces are G and C. Both these letters appear on the force diagram. So through *g* draw a line parallel to GH, and through *c* draw a line parallel to CH. The intersection gives the point *h*.

(8) Finally, the point *j* will be obtained by drawing a line through *h* parallel to HJ and through *d* parallel to DJ. As a test of accuracy the line *ej* should be parallel to EJ.

(9) Measure or calculate the lengths of the lines in the force diagram, and so obtain the forces which they represent. These values are inserted in their appropriate position in the scale diagram at the bottom of Fig. 14·8.

(10) Determine the nature of the forces. The method is illustrated by an example shown in diagram 3, and deals with the forces acting at the apex of the frame. We know the direction of one of these forces, viz. the 800 N force BC. Starting from *b* (in diagram 3), move *down* towards *c* (the 800 N force acts *down*), and continuing in the forward direction, move to the points *h* and *g*, then back to *b*, placing the arrow in position as shown. *These arrows all point the same way round the diagram.* Notice that we are only concerned with the letters *b, c, h, g,* since these are the spaces associated with the apex of the frame. Now transfer the arrows from diagram 3 to diagram 4, so that the four arrows at the apex are pointing in the same direction as in diagram 3. Arrow on *hg* in diagram 3 is down, so that arrow on the central vertical member in diagram 4 (which is HG) is

also down. Repeat for every similar junction point on the frame. As a check *the arrows at opposite ends of the same member must point in opposite directions.* Then, if the arrows point towards each other, the member is in *tension*; if the arrows point away from each other, the member is in *compression.*

(11) Indicate the magnitude and the nature of the various forces on a small-scale diagram as shown at 4.

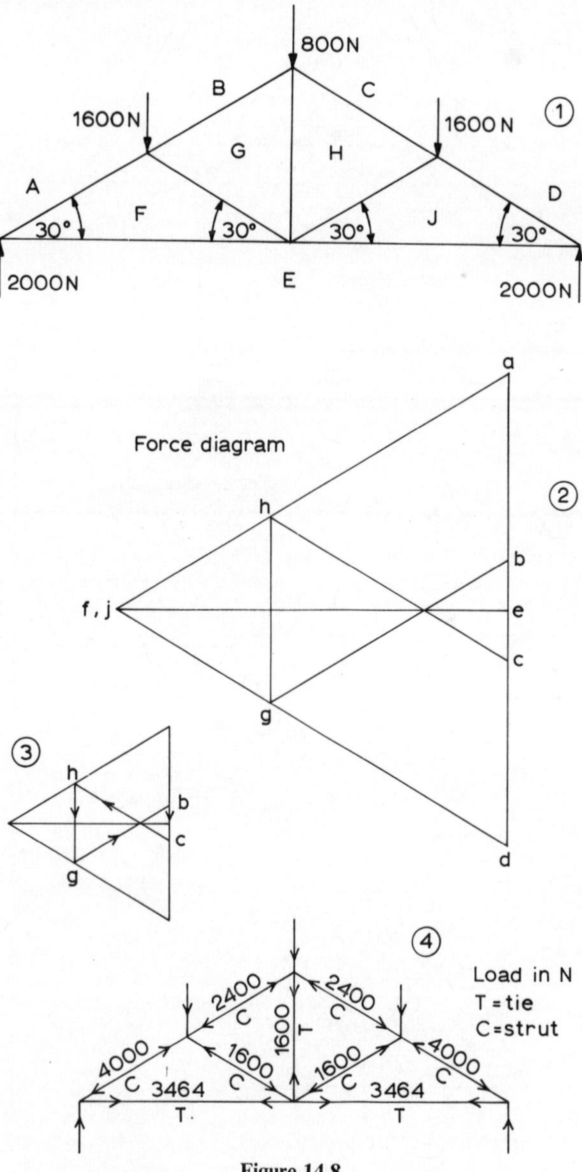

Figure 14.8

8kN
B
12kN
1
F
A
C
E
G
60°
60°
60°
60°
D
8m
8m
2
8kN
12kN
R_a
4m
R_b
12m
a
3
Force diagram
4
a
f
b
e
f
b
g
e
d
5
Load in kN
T= tie
C=strut
6·05
C
10·38
C
1·73
T
1·73
C
13·84
C
5·19
T
6·92
T
9
11
c

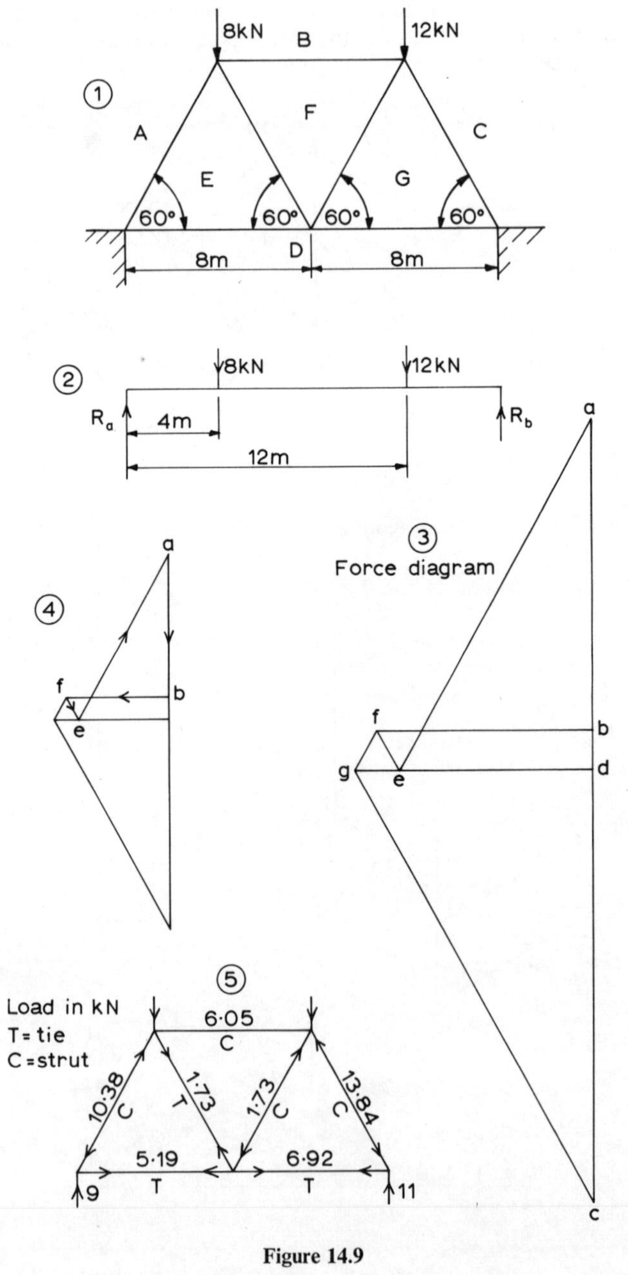

Figure 14.9

Example 14.3

Calculate by moments the reactions at the supports of the truss shown in Fig. 14.9, and find graphically the forces in each of the members.

[N.C.T.E.C.]

METHOD

(1) Draw to scale the space diagram showing the directions of the various members of the frame.

(2) Calculate the reactions in a similar manner to that of a simply supported beam. The triangles are equilateral, so the 8 kN load acts over the mid-point of the base of the left-hand triangle, i.e. at 4 m from the left-hand support. Similarly, the 12 kN load acts at 12 m from the left-hand support. By taking moments, the values of the reactions become $R_a = 9$ kN and $R_b = 11$ kN.

(3) Using Bow's notation, indicate with a capital letter the space between each force and truss member.

(4) Choose a suitable scale for the force diagram, and draw the line *abdc* to represent the vertical external forces and reactions.

(5) The adjoining spaces to space E are A and D. Both these letters are already on the force diagram. Through the point *a* draw a line parallel to AE, and through point *d* draw a line parallel to DE. The intersection gives the point *e* so that *ae* represents the force AE and *de* represents the force DE.

(6) Through *e* draw *ef* parallel to EF, and through *b* draw *bf* parallel to BF The intersection gives point *f*.

(7) Through *f* draw line parallel to FG, and through *c* draw line parallel to CG. The intersection gives point *g*.

(8) Measure or calculate the length of the lines in the force diagram and hence obtain the magnitude of the forces on the members of the frame.

(9) Obtain the direction of the forces acting at the different points of the frame. As an example the forces acting at the top left-hand corner of the frame are dealt with in diagram 4. Starting at *a* follow round the points *a*, *b*, *f* and *e* back to *a*. The arrows should point in the same forward direction with arrow on *ab* pointing *down* since the force AB acts *down*. Transfer these arrows on to the scale diagram 5. Repeat for all other points in the frame. *Remember that arrows at either end of any member must point in opposite directions.* If the arrows point towards each other, the member is in *tension*, if the arrows point away from each other the member is in *compression*.

(10) Indicate the magnitude and nature of the various forces on the scale diagram 5.

Example 14.4

Determine the magnitude and nature of the forces acting in the loaded structure shown in Fig. 14.10.

METHOD

(1) Draw to scale the space diagram showing the directions of the various members of the frame.

(2) Using Bow's notation, indicate with a capital letter the spaces between the forces. The letter C indicates the space between the two wall reactions.

(3) The lower reaction R can be taken as acting in the direction of the lower

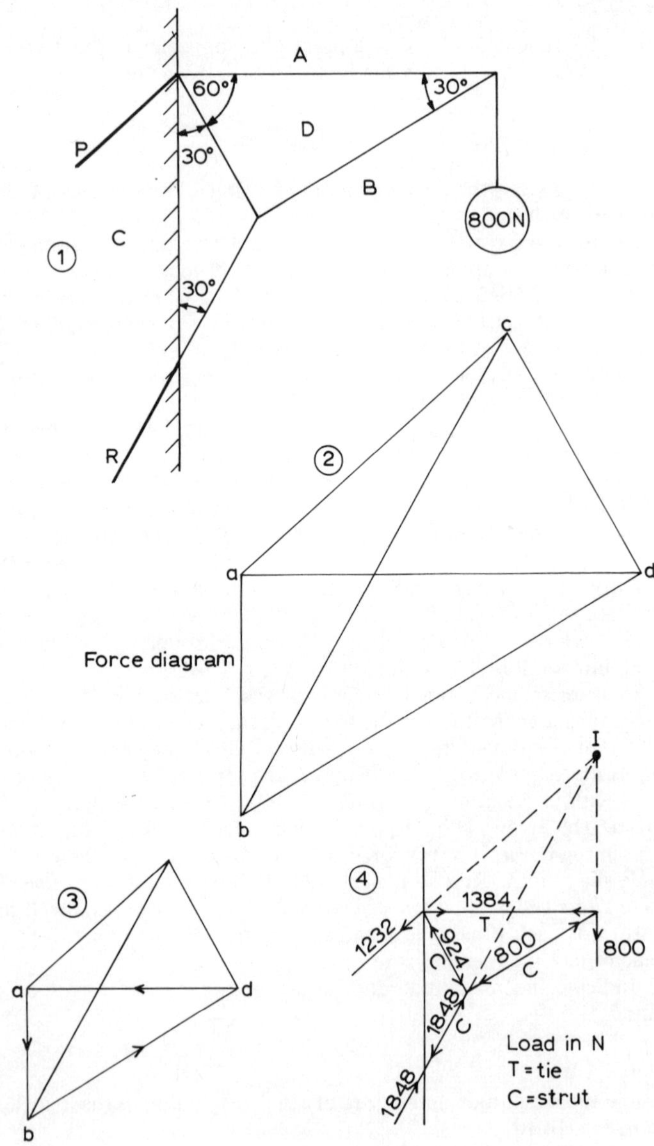

Figure 14.10

member BC as shown. Since all three external forces (800 N, reactions P and R) must pass through one point, the direction of P can be determined. Produce R until it meets the line of action of the 800 N in the point I (diagram 4). Then the line of action of P must also pass through I.

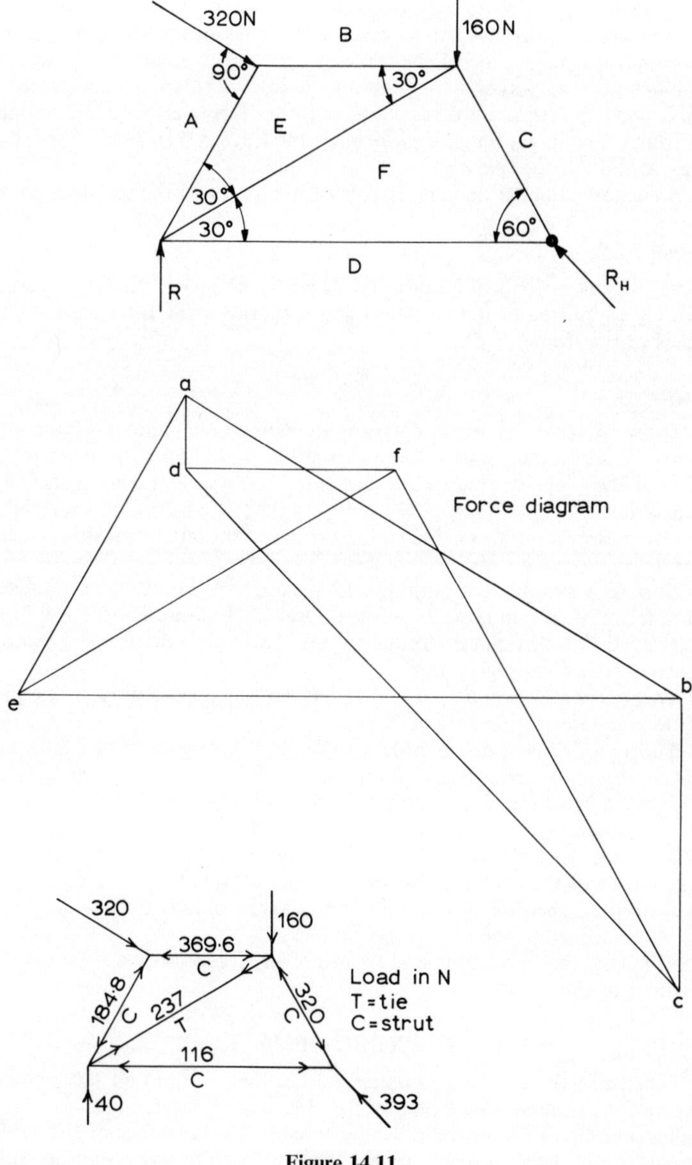

Figure 14.11

(4) Choose a suitable scale for the force diagram 2. Draw *ab* to represent 800 N. Draw *bc* parallel to *R* and *ac* parallel to *P*, intersecting at *c*. Then *bc* represents reaction *R* and *ca* represents reaction *P*.

(5) Through *a* draw *ad* parallel to AD, through *b* draw *bd* parallel to BD;

intersection gives point *d*. As a check, the line *dc* should be parallel to the line CD.

(6) Measure or calculate the lengths of the lines in the force diagram, and hence obtain the magnitude of the forces on the members of the frame.

(7) Directions of forces acting at any point are obtained as before. The example given in diagram 3 deals with the forces acting at the overhanging end of the frame. The direction *ab* is *down* since the force AB is *down*. This indicates the way round the triangle *abd*.

(8) Indicate the magnitude and nature of the forces on the scale diagram 4.

Example 14.5

Fig. 14.11 shows a loaded framework which is hinged at the right-hand end and simply supported at the left-hand end. Determine the forces in the various members of the frame.

METHOD

(1) Draw to scale the space diagram showing the directions of the various members of the frame. The reaction at the left-hand end will be vertical. The reaction at the hinge can be in any direction and will be determined from the force diagram.

(2) Using Bow's notation, indicate with a capital letter the spaces between the forces.

(3) Choose a suitable scale for the force diagram. Draw *ab* to represent in magnitude and direction the 320 N force. Draw *bc* to represent the 160 N force.

(4) Draw a line through *a* parallel to AE. Through *b* draw a line parallel to BE. Intersection gives the point *e*.

(5) Through *e* draw a line parallel to EF, and through *c* draw a line parallel to CF to give intersection at *f*.

(6) Through *f* draw a line parallel to DF, and through *a* draw a line parallel to AD. Intersection gives the point *d*.

(7) Join *cd*. This line gives the magnitude and direction of the reaction at the hinge.

(8) Measure the lengths of the lines on the force diagram and hence obtain the magnitude of the forces in the various members.

(9) Determine the directions of the forces at the different points in the frame in the way outlined in the previous examples.

(10) Indicate the magnitude and nature of the forces on the scale diagram at the bottom of Fig. 14.11.

EXERCISE 14

1–9 Determine the nature and magnitude of the forces in all the members of the various frameworks shown in Fig. 14.12.

10 Find, by graphical construction, the force in each member of the roof truss shown in Fig. 14.13(i), and state whether it is compressive or tensile. Tabulate the forces. [U.E.I.]

11 The structure shown in Fig. 14.13(ii) is hinged at A, and rests on frictionless rollers at B.

(*a*) State direction of the reaction at B.

(*b*) Find graphically the magnitude of the reaction at B, and the magnitude and direction of the reaction at A.

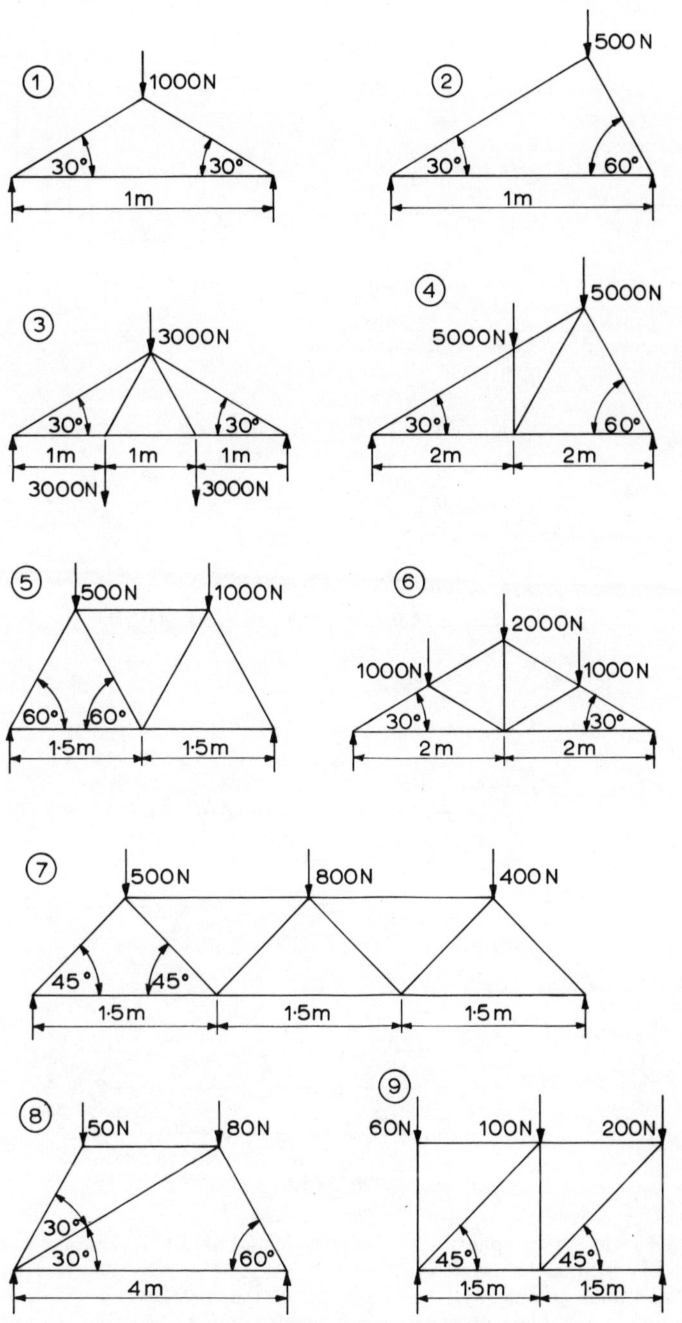

1
1000N
30°
30°
1m
2
500 N
30°
60°
1m
3
3000N
30°
30°
1m
1m
1m
3000N
3000N
4
5000N
5000N
30°
60°
2m
2m
5
500N
1000N
60°
60°
1·5m
1·5m
6
2000N
1000N
1000N
30°
30°
2m
2m
7
500N
800N
400N
45°
45°
1·5m
1·5m
1·5m
8
50N
80N
30°
30°
60°
4m
9
60N
100N
200N
45°
45°
1·5m
1·5m

Figure 14.12

(i)

20kN

0·6m

3m

3m

10m

(ii)

C

5kN

x

A

30°

60°

60°

60°

B

y

7·5m

(iii)

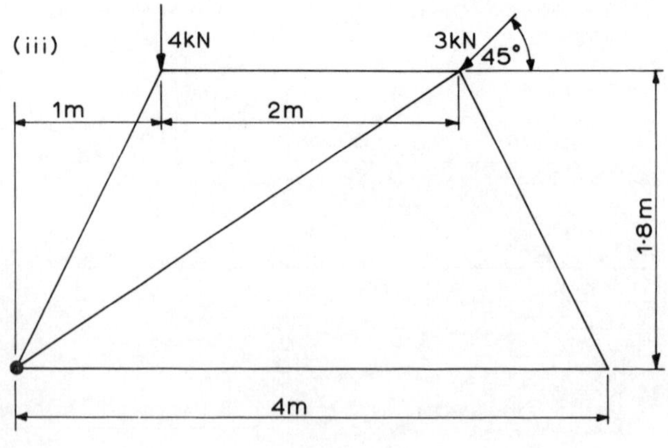

Figure 14.13

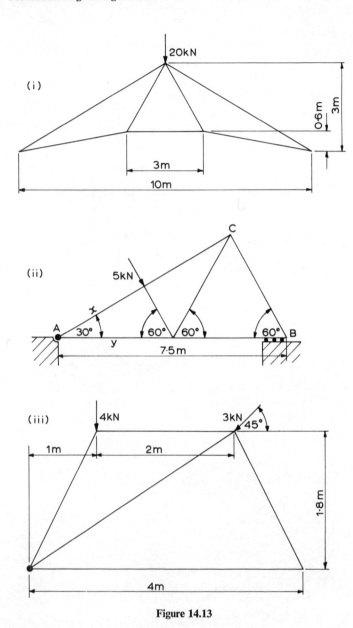

(c) Hence, find graphically the magnitude and nature of the force in each of the members marked x and y. [U.E.I.]

12 Determine, graphically, the forces in all the members of the framework shown in Fig. 14.13(iii) for the loads of 4 kN and 3 kN acting as shown. The framework is hinged at the left-hand end and simply supported at the right-hand end. The magnitudes of the forces should be shown in a table, stating if the member is in tension or compression. [U.E.I.]

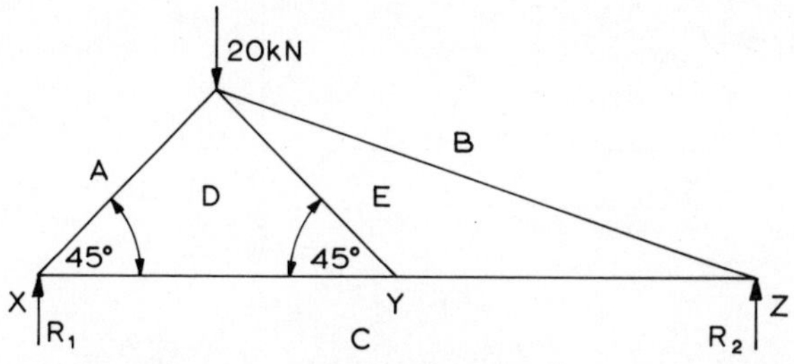

Figure 14.14

13 A jointed roof frame is shown in Fig. 14.14. XY = YZ = 3 m. Find the reactions R_1 and R_2 and draw the force diagram for the members of the frame. Determine the forces in all the members, and tabulate the results, distinguishing between tension and compression members. [U.L.C.I.]

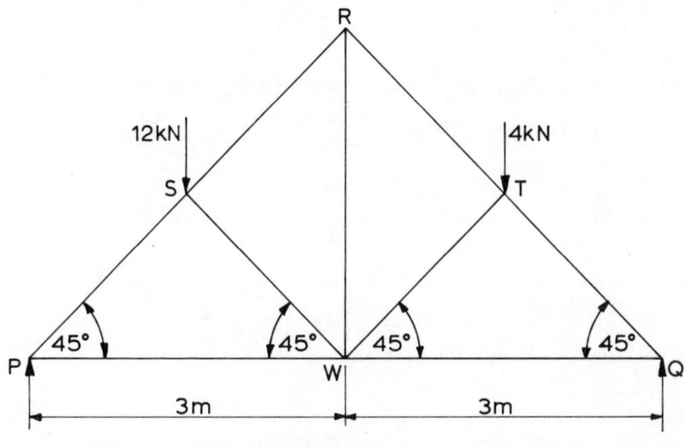

Figure 14.15

14 A simply supported pin-jointed frame is shown in Fig. 14.15. Draw the force diagram and state the magnitude and type of forces in the members PS, SW and RW. Use the following scales: length 1 cm to 50 cm, force diagram, 1 cm to 1 kN. [U.L.C.I.]

15 A simply supported pin-jointed frame is shown in Fig. 14.16. Draw the force diagram and state the magnitude and type of force in the members AF, BC and BF. Use the following scales: length, 1 cm to 50 cm, force diagram, 1 cm to 1 kN. [N.C.T.E.C.]

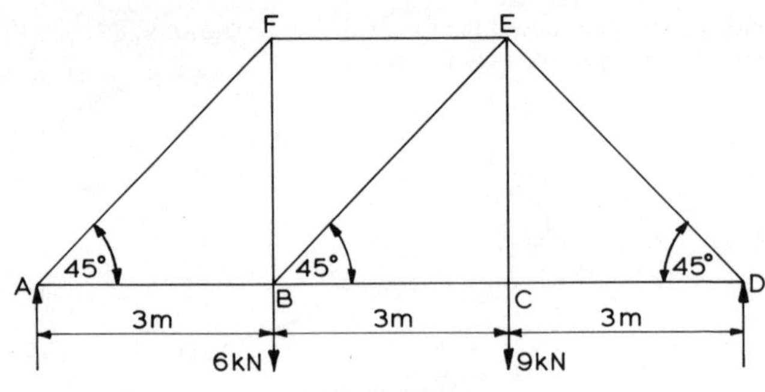

Figure 14.16

16 A cantilever pin-jointed frame is shown in Fig. 14.17. Draw the force diagram and state the magnitude and type of force in the members CE, EB and AB. Use the following scales: space diagram, 1 cm to 50 cm; force diagram, 1 cm to 2 kN.

[N.C.T.E.C.]

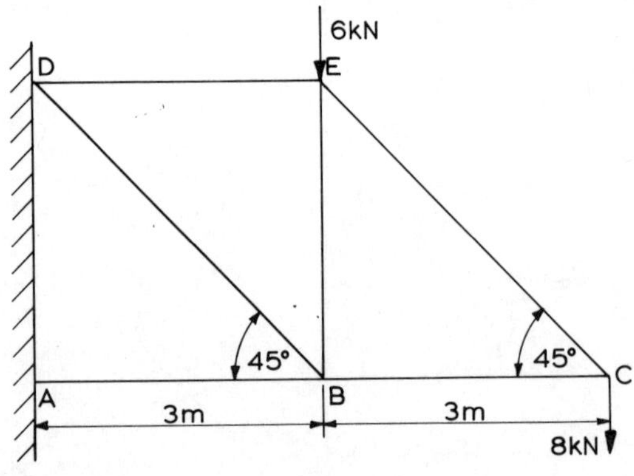

Figure 14.17

Chapter Fifteen
Stress and Strain

First a reminder of two important definitions:

(1) $$\text{Stress} = \frac{\text{force}}{\text{area}}.$$

Stress is measured in newtons per square metre and per square millimetre (N/m^2 and N/mm^2).

(2) $$\text{Strain} = \frac{\text{change of length}}{\text{original length}}.$$

Strain is a number and has no units.

Now there are three fundamental types of stresses:

(*a*) tensile stress,
(*b*) compressive stress,
(*c*) shear stress.

The difference in the nature of these stresses is illustrated in Fig. 15.1.

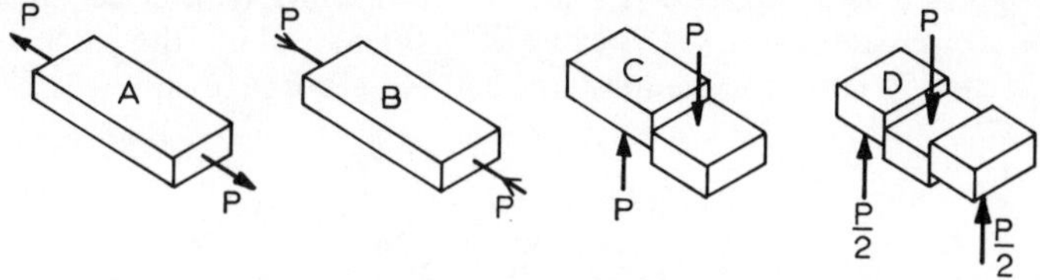

Figure 15.1 *The Three Simple Stresses*
A is in tension
B is in compression
C and D are in shear.

Bar A is subjected to tensile forces, i.e. forces acting along the axis of the bar tending to *pull the bar apart*. Tensile stresses are set up in sections of the bar perpendicular to the direction of the pull.

Bar B is subjected to compressive forces, i.e. forces, acting along the axis of the bar tending to *compress the bar*. Compressive stresses are set up in sections of the bar perpendicular to the direction of the forces.

Bars C and D are subjected to shear forces, i.e. parallel forces not in line, tending to make *one part of the bar slide over the other part*. Bar C is in single shear, while Bar D is in double shear. Shear stresses are set up across the sections parallel to the direction of the applied forces.

In all cases the stress equals $\frac{\text{force}}{\text{area}}$.

For the double shear, two areas are taken into account.

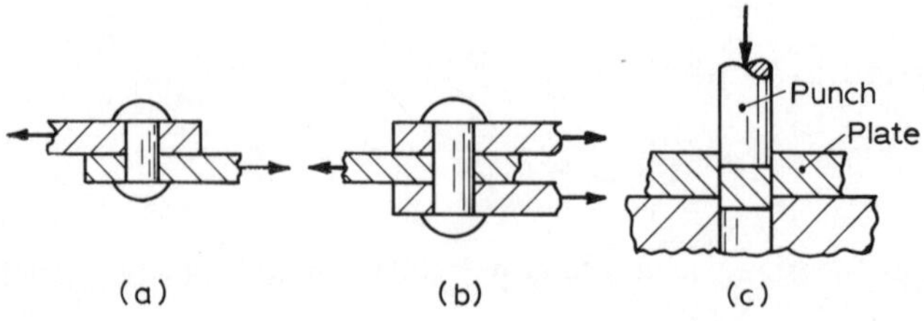

Figure 15.2 *Examples of Shear Stresses*

(*a*) and (*b*) show plates jointed by rivets. The plates are tending to shear the rivet. In (*a*) the rivet is in single shear, and in (*b*) the rivet is in double shear, so that for the same size of rivet the arrangement in (*b*) should be approximately twice as strong as that in (*a*).

(*c*) indicates the action of a punch in punching out a hole in a plate. Normally the punch shears its way through approximately three-quarters of the plate, the remainder of which fails due to tensile forces.

Of course, materials are not equally strong to resist all three types of stresses. For example, cast iron will carry much greater compressive stresses than tensile stresses. A load of 500 N hung from a cast-iron bar to produce tensile stresses will cause it to break, whilst the same size of bar will carry a compressive load of 2250 N before it breaks.

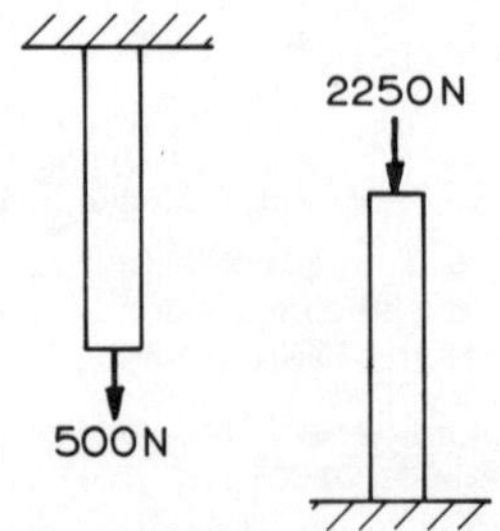

Figure 15.3 *Different Strengths of Materials*

Two identical bars of cast iron are supporting loads. The maximum load which the bar will carry is 500 N if the bar is in tension or 2250 N if the bar is in compression.

Consequently, to design any machine part it will be necessary to know not only the magnitude of the stresses that will be set up, but also their nature.

Relationship between stress and strain

When a load is applied to a test bar, not only are stresses set up, but a change of shape takes place; in other words, the bar is strained. It is either extended or compressed or deflected. If the forces are tensile or compressive, the length of the bar is increased or decreased, and this change of length becomes a measure of the tensile or compressive strain.

$$\text{Tensile or compressive strain} = \frac{\text{change in length}}{\text{original length}}.$$

If the forces are shear forces, then the bar gets pushed out of shape. This is best illustrated by two parallel forces being applied to a block of

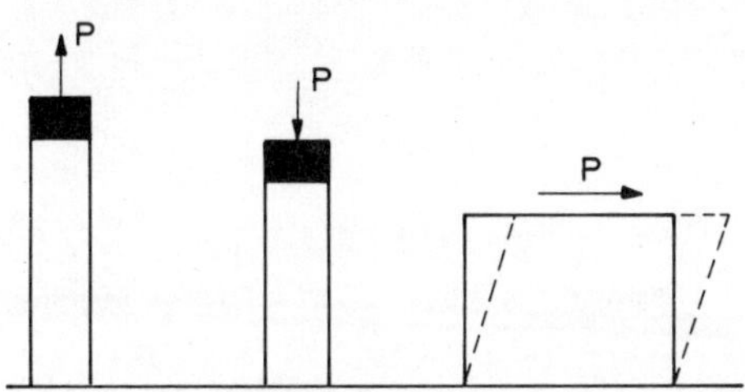

Figure 15.4 *The Three Simple Strains*

The bar on the left increases its length, being subjected to tensile strains.
The bar in the centre decreases its length, being subjected to compressive strains.
The bar on the right is deformed, being subjected to shear strains.

rubber. Actually, only one force is applied to the top surface, and the equal reaction at the bottom provides the other force. The rectangular block is converted into a parallelogram. The movement of the top surface relative to the bottom surface depends upon the distance between the two surfaces (Fig. 15.5).

$$\text{Shear strain} = \frac{x}{AB}$$

$$= \text{angle } \gamma \text{ in radians.}$$

γ is the Greek letter 'gamma'.

Now, if the extension of the bar is kept within the elastic limit (so that when the load is removed the bar will return to its original size), then, according to Hooke's Law:

the strain produced in an elastic material is directly proportional to the stress which produces it.

So that $\frac{\text{stress}}{\text{strain}} = \text{constant}.$

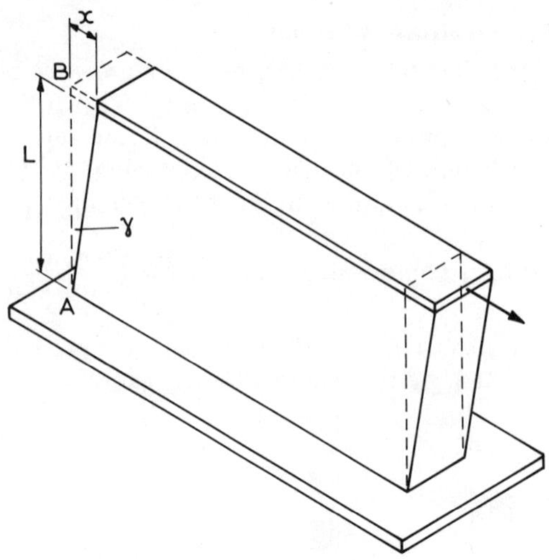

Figure 15.5 *Shear Strain*
The angle γ is a measure of the shear strain.

The constant will depend upon the material and also upon the type of the stress.

For **tensile** or **compressive** stress:

$$\frac{\text{stress}}{\text{strain}} = E, \text{ the modulus of elasticity (sometimes called Young's modulus).}$$

For **shear** stress:

$$\frac{\text{stress}}{\text{strain}} = G, \text{ the modulus of rigidity.}$$

The units for both E and G will normally be newtons per square millimetre or kilonewtons per square millimetre (N/mm^2 or kN/mm^2).

At this stage we are concerned with tensile and compressive stresses and the value of E.

Let W = applied load in N (tensile or compressive),
a = cross-sectional area in mm^2,
L = length in mm,
x = extension in mm,
E = modulus of elasticity in N/mm^2

$$\text{Stress} = \frac{W}{a}\,\text{N/mm}^2$$

$$\text{Strain} = \frac{x}{L} \text{ (note that strain has no units)}$$

$$E = \frac{\text{stress}}{\text{strain}} = \frac{W/a}{x/L}$$

$$E = \frac{WL}{ax} \text{ N/mm}^2$$

Average Values of Elastic Constants

Material	Modulus of elasticity kN/mm^2	Modulus of rigidity kN/mm^2
steel	210	90
copper	90	40
brass	70	37
wood	10	0.6

Load-extension graphs

If we test a specimen of the type shown in Fig. 15.6(a) in a tensile testing machine, we can get an indication of its strength and its behaviour under load. Loads can be applied gradually and the extension of the specimen measured. In the early stages of the experiment we can use a delicate instrument known as an extensometer. As the extension increases we can measure it by means of a pair of steel dividers. We can plot the extension against the load producing it and obtain a load-extension graph. It is, however, more usual to calculate the stress produced by the load and the corresponding strain produced and plot the stress-strain graph. Provided that we use the same cross-sectional area of the specimen for all values of the stress we should obtain a similar diagram for load-extension as for stress-strain. Fig. 15.7 is a typical stress-strain graph for steel.

Between O and A, stress is proportional to strain. The point A is called **the limit of proportionality** or **limit of elasticity,** since beyond this point the metal is no longer elastic and stress is no longer proportional to strain. The metal is in the plastic state. When the load is removed the strain will not disappear, as would be the case up to the elastic limit. The metal is now said to have a permanent set. As the stress is increased, a point is reached when the material begins to yield. The specimen extends without increasing the load; the strain may be increased up to 20 times its previous amount. This is shown by the portion B—C on the graph. The point B at which this occurs is called the yield point. At C the resistance of the bar increases and additional load is required to produce further extension. There now appears to be a marked change in the texture of the metal, especially near the middle portion of the specimen. The surface which was

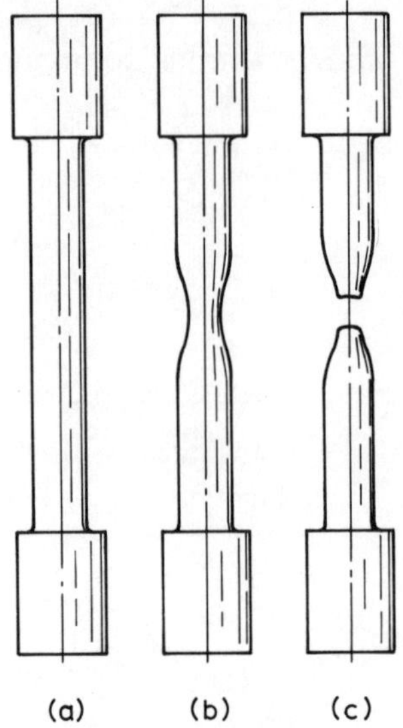

Figure 15.6 *Behaviour of Steel Specimens in Tension*

The specimen approximately 200 mm long and 11 mm in diameter is subjected to an axial pull. As the load increases the specimen increases in length and a 'waist' forms, as shown at (*b*). This reduction in cross-sectional area is followed by a reduction in the load but an increase in the actual stress value. Since the stress in Fig. 15.7 is calculated from load divided by initial cross-sectional area, the stress-strain curve falls off between D and E. Finally the specimen fractures as shown in (*c*).

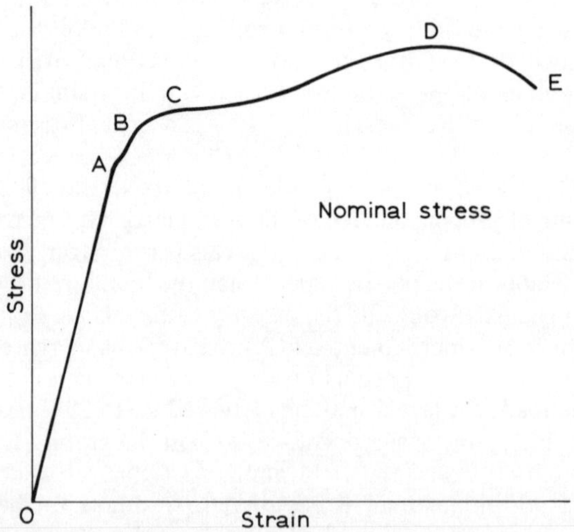

Figure 15.7 *Stress-strain Graph for Steel*

bright becomes dull. There is also a decrease in the diameter of the central portion; a waist begins to form. Due to this change in cross-sectional area, the stress at the waist increases without further increase in the load; in fact, the load is reduced and therefore the graph of the stress calculated on the *original* area will be shown as D to E. The load at D is the maximum load carried by the specimen. Fracture takes place at the point E.

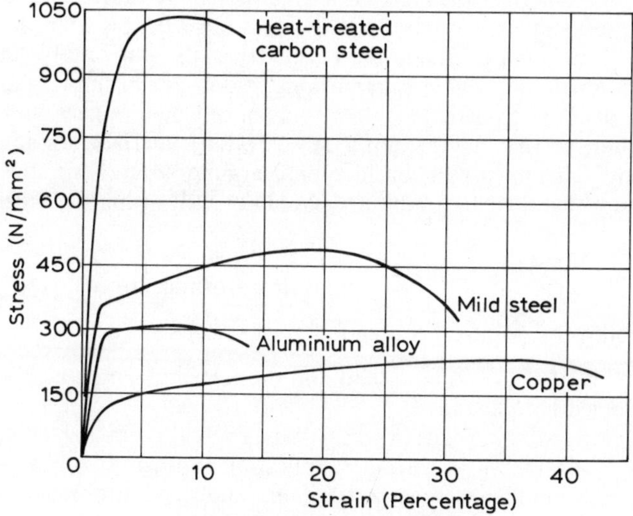

Figure 15.8 *Stress-strain Graphs*

Fig. 15.8 indicates the stress-strain graph for a number of different materials. Graphs of this type are a great help to the designer. You can see that the heat-treated carbon steel will stand much more stress than the mild steel. You can also see that the carbon steel quickly breaks after the elastic limit has been passed. This is quite different from the copper. Copper continues to extend considerably after passing the elastic limit before it finally breaks.

Tensile strength

The tensile strength of a material is an indication of the maximum tensile stress which can be carried by the material without causing the material to break.

$$\text{Tensile strength} = \frac{\text{maximum load}}{\text{original area}} \text{ N/mm}^2.$$

The tensile strength is of importance in the design of machines and structures. The various parts must not be stressed beyond the allowable working stress, and this must be below the elastic limit of the material. The designer will then ensure that the part remains elastic, i.e. that it returns to its original dimensions on the removal of the stress.

Factor of safety

The load which any member of a machine carries is called the working load, and the stress produced by this load is the 'working' or 'allowable' stress. Obviously the working load must be less than the ultimate breaking load of the member, i.e. the working stress must be less than the tensile strength. It is usual to divide the tensile strength by a suitable factor of safety to get the allowable safe working stress. For example, the tensile strength of a material may be 450 N/mm^2, but the designer may design the machine so that the maximum stress in the material does not exceed 80 N/mm^2. There is thus a considerable margin of safety. The stresses and strains to which the bar is subjected should be well within the elastic limit.

$$\text{Factor of safety} = \frac{\text{tensile strength}}{\text{allowable working stress}}.$$

The following points affect the factor of safety.

1 The value of the maximum load and the certainty with which it can be determined beforehand.
2 The nature of the load: whether it is gradually applied or suddenly applied; whether it is a single application (known as 'dead' load) or repeated a number of times per minute (known as 'live' load); whether it is acting in one direction or suddenly reversing direction, i.e. producing alternately tensile and compressive stress in the material.
3 The reliability of the material and the certainty with which its strength can be stated.
4 The effect of corrosion or wear on the dimensions of the part.
5 Possible errors of workmanship in the manufacture of the component.
6 The consequences of a breakdown.

The value of the factor of safety varies from 3, in the case of steel carrying a dead load, to 20, in the case of timber carrying a shock or suddenly applied load.

Strain energy

When a force is applied to an elastic material, such as a spring or a bar of steel, deflection or extension takes place. The force therefore moves through a distance and work is done. This work is stored in the material as strain energy.

Now, the force W acting on the spring is not constant throughout the whole of the extension, but varies from zero to the maximum value W, even though the loading may occupy only a short interval of time. The load-extension diagram is shown in Fig. 15.9.

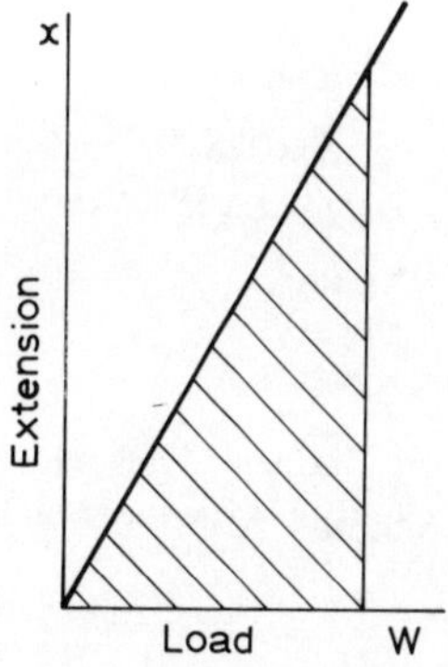

Figure 15.9 *Strain Energy*
The shaded area of the diagram represents the work done in straining a solid or spring. This work is stored in the spring as strain energy, and, provided the elastic limit has not been exceeded, most of the energy will be given up when the spring is released.

The work done in straining the spring

$$= \text{area under graph}$$
$$= \tfrac{1}{2}Wx \text{ Nm.}$$

This is the strain energy stored in the spring. When the spring is allowed to return to its normal length, it is capable of doing work to the extent of $\frac{1}{2}Wx$ Nm. The actual value will be slightly less than this, due to internal friction in the material.

In dealing with the strain energy stored in a spring, it is often useful to know the stiffness of the spring. By stiffness of a spring is meant the force required to cause unit deflection or extension. If a spring stiffness is expressed as 2·5 N/mm, a force of 2·5 N will be required to produce a deflection of one millimetre.

Example 15.1

A metal bar of length 250 mm and cross-sectional area 150 mm² is gradually loaded in tension up to the elastic limit. From the load-extension graph it is found that a load of 25 kN produces an extension of 0·2 mm. Find Young's modulus for the metal. [N.C.T.E.C.]

$$\text{Stress under load of 25 kN} = \frac{\text{load}}{\text{area}}$$
$$= \frac{25}{150} \text{ kN/mm}^2$$
$$= 166{\cdot}7 \text{ N/mm}^2.$$
$$\text{Strain at this load} = \frac{\text{extension}}{\text{original length}}$$
$$= \frac{0{\cdot}2}{250}$$
$$= 0{\cdot}0008$$
$$\text{Young's modulus} = \frac{\text{stress}}{\text{strain}}$$

$$= \frac{166{\cdot}7}{0{\cdot}0008}$$
$$= 208\ 375\ \text{N/mm}^2$$
$$= 208{\cdot}4\ \text{kN/mm}^2.$$

Young's modulus for the metal is 208·4 kN/mm².

Example 15.2

Determine the extension of a rod 15 mm in diameter, 200 mm long when subjected to a load of 20 kN. Modulus of elasticity for the material is 200 kN/mm².

$$\text{Area of rod} = \frac{\pi}{4} \times 15^2$$
$$= 177\ \text{mm}^2$$
$$\text{Stress} = \frac{\text{load}}{\text{area}} = \frac{20 \times 1000}{177}\ \text{N/mm}^2$$
$$= 113\ \text{N/mm}^2$$
$$\frac{\text{Stress}}{\text{Strain}} = E$$
$$\text{Strain} = \frac{\text{stress}}{E}$$
$$= \frac{113}{200 \times 1000}$$
$$= 0{\cdot}000\ 565$$
$$\text{Strain} = \frac{\text{extension}}{\text{original length}}$$
$$\text{Extension} = \text{strain} \times \text{original length}$$
$$= 0{\cdot}000\ 565 \times 200$$
$$= 0{\cdot}113\ \text{mm}.$$

Extension of rod is 0·113 mm.

Example 15.3

If the tensile strength of a certain steel is 400 N/mm², what will be the maximum load which a rod 2·5 m long and 40 mm in diameter can carry with a factor of safety of 5? What will then be the extension if $E = 200\ \text{kN/mm}^2$?

$$\text{Allowable working stress} = \frac{\text{Tensile strength}}{\text{Factor of safety}}$$
$$= \frac{400}{5} = 80\ \text{N/mm}^2$$
$$\text{Working load} = \text{area} \times \text{stress}$$
$$= \frac{\pi}{4} \times 40^2 \times 80 \times 1000\ \text{kN}$$

$$= 100 \text{ kN}$$

$$\text{Strain} = \frac{\text{stress}}{E}$$

$$= \frac{80}{1000 \times 200}$$

$$= 0{\cdot}0004$$

$$\text{Extension} = \text{strain} \times \text{original length}$$

$$= 0{\cdot}0004 \times 2{\cdot}5 \times 1000$$

$$= 1 \text{ mm}$$

The maximum working load is 100 kN, giving an extension of 1 mm.

Example 15.4

Fig. 15.10 shows a coupling-rod connection. If the rods carry a load of 150 kN, determine suitable diameters for the rod and the pin. Allowable tensile stress for the rod, 70 N/mm². Allowable shear stress for the pin, 60 N/mm².

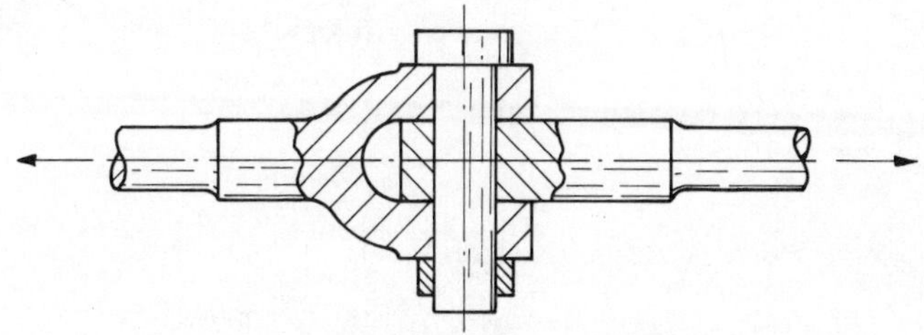

Figure 15.10 *Simple Coupling for Rods*

The pin which holds the rods rogether is subjected to double shear when the rods are loaded as shown.

$$\text{Load carried by rod} = 150 \text{ kN}$$

$$\text{Cross-section area of rod} = \frac{\text{load}}{\text{stress}}$$

$$= \frac{150 \times 1000}{70} \text{ mm}^2$$

$$= 2140 \text{ mm}^2.$$

If d is the diameter of the rod in mm, then

$$\frac{\pi}{4}d^2 = 2140$$

$$d = 52 \text{ mm}.$$

The pin is in double shear.

If D is the diameter of the pin in mm, then total area resisting shear is $2 \times \frac{\pi}{4}D^2$.

$$\text{Load} = \text{stress} \times \text{area}$$

$$150 \times 1000 = 60 \times 2 \times \frac{\pi}{4}D^2$$

whence $D = 39{\cdot}9$ mm.

Suitable sizes would be 52 mm for the rod and 40 mm for the pin.

Example 15.5

Calculate the force required to punch a hole 15 mm in diameter in a plate 10 mm thick if the shear strength of the material is 350 N/mm². What would be the compressive stress in the punch?

In this case, the area resisting shear is equal to circumference of hole × thickness of plate.

$$\begin{aligned} \text{Area resisting} &= 15\pi \times 10 \\ &= 472 \text{ mm}^2 \\ \text{Force} &= \text{stress} \times \text{area} \\ &= 350 \times 472 \text{ N} \\ &= 165 \text{ kN} \end{aligned}$$

$$\begin{aligned} \text{Cross-sectional area of punch} &= \frac{\pi}{4} \times 15^2 \\ &= 177 \text{ mm}^2 \\ \text{Compressive stress} &= \frac{\text{load}}{\text{area}} \\ &= \frac{165 \times 1000}{177} \\ &= 932 \text{ N/mm}^2. \end{aligned}$$

Force to punch hole is 165 kN and compressive stress in punch is 932 N/mm².

Example 15.6

A spring whose stiffness is 10·5 N/mm is compressed a distance of 50 mm. How much strain energy is stored in the spring? If the spring is used to open a valve whose mass is 200 g, what will be the maximum velocity of the valve assuming that it acts in a horizontal direction and that friction can be neglected?

$$\begin{aligned} \text{Load to compress spring} &= \text{stiffness} \times \text{compression} \\ &= 10{\cdot}5 \times 50 \\ &= 525 \text{ N} \\ \text{Strain energy} &= \tfrac{1}{2} \text{ force} \times \text{distance} \\ &= \tfrac{1}{2} \times 525 \times \frac{50}{1000} \\ &= 13{\cdot}1 \text{ Nm} \\ &= 13{\cdot}1 \text{ joules.} \end{aligned}$$

Let v m/s be maximum velocity of valve.

$$\text{Mass of valve} = 200 \text{ g} = 0{\cdot}2 \text{ kg}$$

$$\text{Kinetic energy at max. speed} = \tfrac{1}{2}\ \text{mass} \times \text{velocity}^2$$
$$= \tfrac{1}{2} \times 0{\cdot}2 \times v^2$$

By the principle of the conservation of energy, the strain energy of the spring must equal the kinetic energy of the valve (since no work is done against friction).

$$\therefore \quad \tfrac{1}{2} \times 0{\cdot}2 \times v^2 = 13{\cdot}1$$
$$v^2 = 131{\cdot}0$$
$$v = 11{\cdot}4\ \text{m/s.}$$

Strain energy stored in spring is 13·1 joules and maximum velocity of the valve is 11·4 m/s.

EXERCISE 15

1 A bar of steel 500 mm long has a cross-sectional area of 500 mm² and is subjected to a direct pull of 20 kN. Determine the stress in the bar and the elongation if E is 200 kN/mm².

2 If the tensile strength of a steel wire is 750 N/mm², what load could be safely carried by the wire of diameter 2·5 mm? Use a factor of safety of 5. If the wire is 6 m long and the observed extension is 4·8 mm, what is the value of the modulus of elasticity?

3 A round brass rod in tension carries a load of 40 kN. Determine its diameter if the stress is not to exceed 40 N/mm². What will be the extension of the rod if the length is 2 m, and the modulus of elasticity for brass is 70 kN/mm²?

4 The coupling for a tie-rod carrying a load of 100 kN is similar to that shown in Fig. 15.10. If the allowable tensile stress for the rod is 84 N/mm² and the allowable shear stress for the pin is 55 N/mm², calculate the diameter of the rod and the pin.

5 If the ultimate shear stress for a boiler-plate material is 400 N/mm², determine the force required to punch a hole 20 mm in diameter in a plate 25 mm thick. What would be the compressive stress in the punch?

6 The extension of a bar 6 m long and 50 mm by 25 mm section is to be limited to 2·5 mm. Determine the maximum load which the bar can carry. $E =$ 200 kN/mm².

7 A load of 2000 N is applied to a steel wire 200 mm long and 2·5 mm in diameter. If E is 200 kN/mm², determine the extension of the wire. How much strain energy is stored in the wire?

8 A spring whose stiffness is 12 N/mm is compressed 150 mm. What load must have been applied? How much energy has been stored in the wire?

9 A metal bar of length 250 mm and cross-sectional area 150 mm² is gradually loaded in tension up to the elastic limit. From the load-extension graph it is found that a load of 25 kN produces an extension of 0·2 mm. Find Young's modulus for the metal. [N.C.T.E.C.]

10 With the aid of sketches, distinguish between tensile stress and shearing stress.

Calculate the force required to punch a hole 20 mm in diameter in a plate 10 mm thick if the shear strength of the material is 500 N/mm². [N.C.T.E.C.]

11 A column in a structure may be either cast in iron or rolled in steel, but a tie-bar would be unsuitable if made of cast iron. Explain briefly why this is so.

A short cast-iron column is to be a hollow cylinder, 150 mm in external diameter, and is to support an axial load of 400 kN. What would be a suitable internal diameter for this column if the working stress is not to exceed 100 N/mm²? [N.C.T.E.C.]

12 Define the term 'elastic limit.'

A steel tie-rod in a structure is 25 mm in diameter and 10 m long, and it is subjected to a pull of 80 kN. Calculate (*a*) the elongation, (*b*) the factor of safety. E for the material is 200 kN/mm² and its tensile strength is 450 N/mm². [N.C.T.E.C.]

13 A vertical steel wire 5 m long and 2 mm in diameter carries a load of 500 N. Calculate the extension of the wire if $E = 200$ kN/mm². [N.C.T.E.C.]

14 Define strain, stress, and modulus of elasticity.

A steel tie-bar, 5 m long and 65 mm by 15 mm cross section, withstands a pull of 125 kN. Determine the tensile stress on the section and the increase in length of the bar under this pull. Take Young's modulus, $E = 200$ kN/mm². [N.C.T.E.C.]

15 A mild-steel tie-bar, of circular cross section, lengthens 1·5 mm under a steady pull of 75 kN. The original dimensions of the bar were 5 m long and 30 mm in diameter. Find the intensity of tensile stress in the bar, and the value of the modulus of elasticity. [N.C.T.E.C.]

16 Within certain limits, the amount of extension of a spring is directly proportional to the load applied. It is found that a force of 50 N is required to stretch a spiral spring by 4·5 mm. Gradually increasing loads are applied in succession as follows: 50 N, 100 N, 150 N, 200 N, 250 N. Draw a graph having loads as ordinates and extensions as abscissae. Find the work done in joules in stretching the spring (*a*) through the first 13·5 mm, (*b*) during the increase in load from 150 N to 250 N, and (*c*) state the values obtained in cases (*a*) and (*b*) as percentages of the *total* work done during the experiment. [N.C.T.E.C.]

17 A British Standard specification for structural steel limits the working stress for this material to 150 N/mm².

(*a*) If a tie-rod of this material has to withstand a total pull of 170 kN, determine a suitable diameter for the tie-rod.

(*b*) Calculate the extension of the rod under this load if the rod is 3 m long. $E = 200$ kN/mm². [U.E.I.]

18 A test on a duralumin test piece 150 mm² in cross-sectional area and 50 mm long showed that it followed Hooke's Law up to a load of 38 kN with an extension of 1·6 mm for that load.

Calculate the stress at this load and the value of Young's modulus for duralumin. Use your results to find the load that can be carried by a duralumin tension member 425 mm² in cross-sectional area and 325 mm long if the maximum permissible extension is 7·5 mm. [U.E.I.]

19 A mild-steel rod, 1350 mm² in cross-sectional area, and 4 m long, at a temperature of 75° C is allowed to cool to 15° C. If it could contract freely, the rod would shorten by an amount 4000 (75 − 15) × 0·000 011 2 mm. It is, however, gripped at its ends and the contraction in length is only 0·5 mm.

Calculate the intensity of the stress set up and the total force exerted by the rod.

Modulus of elasticity = 200 kN/mm². [U.E.I.]

20 (*a*) Explain clearly what is meant by the terms 'stress,' 'strain,' 'modulus of elasticity.'

(*b*) A mild-steel rod has to carry an axial pull of 48 kN. The maximum permissible stress is 125 N/mm². Calculate a suitable diameter for the rod. [U.E.I.]

21 (*a*) Explain the meanings of the terms 'stress,' 'strain,' 'modulus of elasticity,' and 'factor of safety.'

(*b*) A tie-bar 50 mm in diameter and 3 m long is made of steel whose tensile strength is 450 N/mm². Find the safe load that may be carried by the tie-bar if the factor of safety is 4.

(*c*) Find the extension of the bar under this load, if $E = 200$ kN/mm². [U.E.I.]

22 A spring whose stiffness is 10 N/mm is compressed a distance of 80 mm.

(*a*) Draw a diagram the area of which represents to some scale the energy stored in the spring, and state the scale.

(*b*) From your diagram find the energy stored in the spring.

(*c*) If the compressed spring is used to open a valve weighing 2·5 N horizontally, calculate the maximum speed of opening of the valve. [U.E.I.]

23 Define: (*a*) yield stress, (*b*) tensile strength, and (*c*) factor of safety.

Tests on a sample of mild steel give a tensile strength of 450 N/mm², and for a stress of 60 N/mm² a strain of 0·000 3 is produced. If a tie-bar of this steel is 2 m long and subjected to a load of 20 kN, determine: (*a*) a suitable cross-sectional area allowing a factor of safety of 5, and (*b*) the extension produced by the load. [N.C.T.E.C.]

24 A mild steel test specimen 25 mm in diameter was subjected to a gradually increasing tensile load in a testing machine. The elongation on a length of 250 mm was measured and the following results obtained:

Load (kN)	20	40	60	80	100	
Elong. (mm)	0·048	0·097	0·142	0·196	0·241	
Load (kN)	120	140	160	170	180	190
Elong. (mm)	0·287	0·343	0·39	0·42	0·46	0·52

Plot a graph of load against extension and from it determine the value of Young's modulus for the material and the stress at the limit of proportionality.

25 Define stress and strain.

A tie-bar is to be designed to carry a load of 450 kN, with a factor of safety of 4.

The material to be used has an ultimate tensile stress of 500 N/mm², Young's modulus 115 kN/mm².

Determine:

(*a*) a suitable diameter for a solid circular section,

(*b*) the amount by which the tie-bar will stretch, assuming its length to be 3·5 m. [U.E.I.]

26 (*a*) Define (i) modulus of elasticity and (ii) yield point.

(*b*) A mild steel specimen of diameter 15 mm was found to have an extension of 0·2 mm on a 200 mm gauge length when a load of 35 kN was applied. Determine Young's modulus for the material assuming the load to lie within the limit of proportionality. If a load of 70 kN produced an extension of 0·9 mm on the 200 mm gauge length, determine whether the stress would still be within the limit of proportionality. [N.C.T.E.C.]

Chapter Sixteen
Measurement of Pressure

The idea of pressure is a very important one for the engineer. He needs to know for instance, the magnitude of the pressure exerted by reservoir water on the wall of the dam, or the maximum pressure which a gas container will withstand before it explodes, or the pressures produced on a retaining wall by a quantity of soil behind it. What factors are involved in the measurement and assessment of these pressures?

Pressure is defined as *force per unit area.* Unit pressure would be produced by a force of one newton acting uniformly over an area of one square metre and would be expressed as 1 newton per metre^2 (1 N/m^2). The name 'pascal' is increasingly being used for this unit of pressure (1 Pa = 1 N/m^2).

There is an alternative unit which is acceptable, at least for the time being, and this is a 'bar', which is used extensively in meteorology (atmospheric pressures in millibars are familiar). The bar is a much larger unit than the one defined above. It has certain advantages particularly when dealing with the pressure of gases

$$1 \text{ bar} = 10^5 \text{ newtons per metre}^2 \text{ or } 100 \text{ kN/m}^2$$

The bar is approximately equal to the pressure exerted by the atmosphere under normal conditions, at sea level. This pressure, incidentally is called 'one atmosphere'.

It is the pressure of the atmosphere with which we are most familiar, although we are not necessarily aware of it. We have become used to living and working in the environment of atmospheric pressure. If we have to work underwater or travel in aircraft, precautions have to be taken to allow these activities to be carried out with the minimum discomfort.

Measurement of atmospheric pressure: the barometer

If a closed tube containing a liquid has its open end inserted into a container of the same liquid, as shown in Fig. 16.1, the level of the liquid reaches a position of equilibrium at a height h metres above the level of the liquid in the container.

Let the cross-sectional area of the tube be a m^2

Then the volume of liquid in the column is ah m^3.

If the density of the liquid is ρ kg/m^3, then its mass will be $ah\rho$ kg and its weight will be $ah\rho g$ newtons. (ρ is the Greek letter 'rho'.)

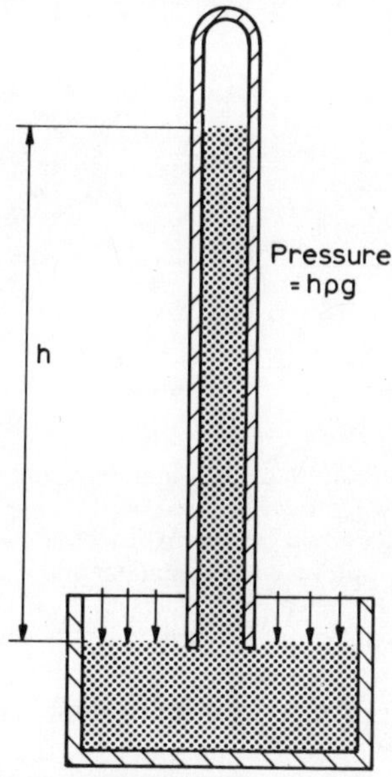

Figure 16.1 *Barometer*

$$\text{Pressure exerted by this column at its base} = \frac{\text{force}}{\text{area}}$$

$$= \frac{ah\rho g}{a}$$

$$= h\rho g.$$

This pressure is balanced by the pressure of the atmosphere acting on the liquid in the container. Hence, the height of the column of liquid is a measure of the pressure of the atmosphere.

If the liquid is water, then ρ is 10^3 kilograms per cubic metre, while the corresponding value for mercury is $13{\cdot}6 \times 10^3$ kg/m^3.

Under normal conditions at sea level, the height of a column of mercury supported by the pressure of the atmosphere is 0·76 metres. This is equivalent to a pressure of $0{\cdot}76 \times 13{\cdot}6 \times 10^3 \times 9{\cdot}81 = 101{\cdot}4$ kN/m^2, or $1{\cdot}014 \times 10^5$ N/m^2, which is 1·014 bar. Note that 1 bar is therefore approximately equal to the normal pressure of the atmosphere at sea level.

Pressure exerted by a head of liquid

By head of liquid we mean the vertical distance from the point considered to the surface of the liquid. Glass vases of different shapes are arranged to fit over a circular plate at their base (Fig. 16.2). This plate is pivoted and provided with a counterbalance weight. By adjusting the position of this weight, we can balance the force on the cover-plate due to the pressure of the water. The vases can be interchanged and the experiment repeated. In

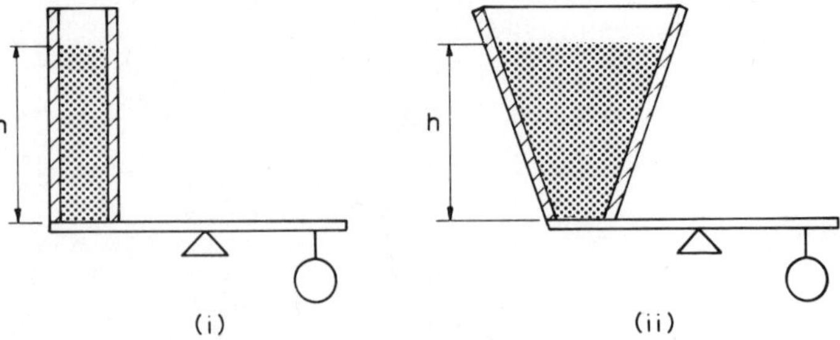

Figure 16.2 *Pressure Due to Head of Liquid*

The base of each vessel is kept in place by a counterbalance weight. The same weight is required in each case, showing that the force on the base of each vessel is the same. The pressure produced by the liquid on the base of the vessel is dependent upon the head of liquid h. The fact that vessel (ii) contains more liquid does not affect the pressure on the base. Note in each case the vessel is supported separately, i.e. the lever supports only the liquid—not the vessel.

all cases it will be found that the position of the balance weight is the same, showing that the force exerted by the water is the same in each case. We should no doubt be inclined to say that the water in the vase (*b*) would exert the greater force, since this vase contains the greater quantity of water. But the pressure exerted by any liquid at a point is due entirely to the height of liquid above the point and to the density of the liquid.

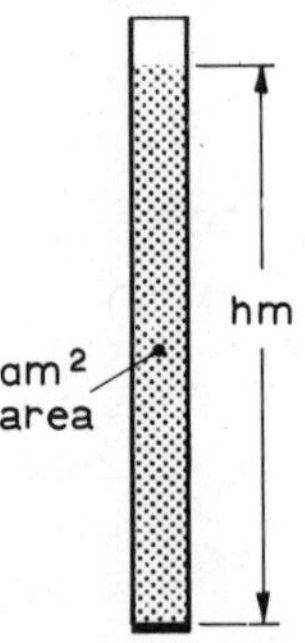

Figure 16.3 *Pressure exerted by a Column of Liquid*

Consider a column of water h m high and a m^2 in cross-sectional area. A cover is placed over the base of the column, and we require to know the force on the cover. The cover is supporting the whole of the column of water, therefore the force on the cover will be the weight of the water.

$$\begin{aligned}\text{Volume of water in column} &= \text{height} \times \text{area of base} \\ &= ha \text{ m}^3\end{aligned}$$

1 m^3 of water has a mass of 1000 kg.

$\therefore$ 1 m^3 of water weighs 1000 g N.

$$\text{Weight of this column of water} = 1000gha \text{ N}$$
$$\text{Force on cover-plate} = 1000gha \text{ N}$$

But area of cover-plate is equal to a m^2

$$\text{Pressure on plate} = \frac{\text{force}}{\text{area}}$$
$$= 1000gh \text{ N/m}^2 \ (g = 9{\cdot}81 \text{ m/s}^2).$$

The pressure exerted by a column of water is 9·81 kN/m^2 for every metre in height of the column.

If a liquid other than water is considered, we can determine the pressure exerted at any depth if the density, ρ, of the liquid is known.

Pressure exerted by h metres of liquid, density ρ, $= 9{\cdot}81 \times \rho \times h$ N/m^2.
(N.B. ρ must be expressed in kg/m^3.)

Absolute pressure

It is important to note that there is another force on the plate in addition to that provided by the head of liquid. The atmospheric pressure acting on the liquid at the top of the column is transferred throughout the length of the column to the plate at the base. In effect, the actual pressure on the plate is equal to:

Pressure due to head of liquid + pressure due to atmosphere.

This is the basis of the important relationship:

absolute pressure = pressure due to head of liquid + pressure of atmosphere

NOTE. When considering the pressure of the atmosphere in the case illustrated in Fig. 16.1, the upper end of the tube was considered closed. Therefore no pressure could be exerted by the atmosphere on the top of the column of liquid.

Table of Density for some Liquids

Substance	Density kg/m^3
alcohol	790
mercury	13 600
oil	870
paraffin	800
water	1 000
sea water	1 025

Example 16.1

A tank whose base measures 8 m by 10 m contains water to a depth of 5 m.

Calculate the pressure on the base in kN/m², and also the total force exerted by the water on the base.

Pressure on base is dependent upon the head of water only.

$$\begin{aligned}\text{Pressure due to 5 m head} &= 10^3 \times 9{\cdot}81 \times 5\\ &= 49\,050 \text{ N/m}^2\\ &= 49{\cdot}05 \text{ kN/m}^2\\ \text{Total force on base} &= \text{pressure} \times \text{area}\\ &= 49{\cdot}05 \times 8 \times 10 \text{ kN}\\ &= 3924 \text{ kN}\end{aligned}$$

The pressure on the base is 49 kN/m² and the total force on the base is 3924 kN.

Example 16.2

A petrol storage tank is 12 m by 15 m by 8 m high, and contains petrol to a depth of 7 m. If the density of the petrol is 700 kg/m³, determine the pressure on the base in N/m² and also the total force exerted on the base.

Pressure on base due to 7 m head of liquid whose density is 700 kg/m³.

$$\begin{aligned} &= 9{\cdot}81 \times 700 \times 7\\ &= 48\,069 \text{ N/m}^2.\\ \text{Total force on base} &= \text{pressure} \times \text{area}\\ &= 48\,069 \times 12 \times 15 \text{ N}\\ &= 8652 \text{ kN}\end{aligned}$$

The pressure on the base is 48 069 N/m² and the total force on the base is 8652 kN.

There are two other important principles with which we should be familiar in connection with pressure exerted by a liquid.

(1) **The pressure exerted by a liquid is always perpendicular to the surface of the containing vessel.**

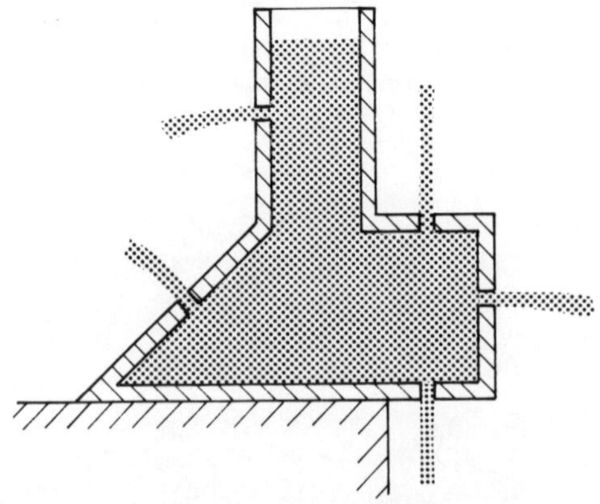

Figure 16.4 *Pressure of a Liquid*

The pressure exerted by a liquid is always perpendicular to the side of the enclosing vessel.

This is something which can be demonstrated experimentally rather than proved mathematically. If we have a vessel similar to that shown in Fig. 16.4 which can be filled with water, then, by removing the covers at the various outlets, a jet of water emerges from each hole in a direction perpendicular to the surface of the vessel at the outlet; thus, a hole at the top produces a vertically upward jet, one at the vertical side produces a horizontal jet, and one in the base produces a vertically downward jet. Now these directions are determined by the direction of the water pressure behind the hole. Hence, it can be taken that liquid pressure is always perpendicular to the surface of the containing vessel.

(2) **The pressure of a liquid inside a closed vessel is constant throughout the vessel.**

Consider a closed vessel fitted with a number of pistons as shown in Fig. 16.5. By applying a force to one of the pistons, we subject the liquid to a pressure given by force ÷ area of piston. Due to the fact that liquids are practically incompressible (this being one of the fundamental differ-

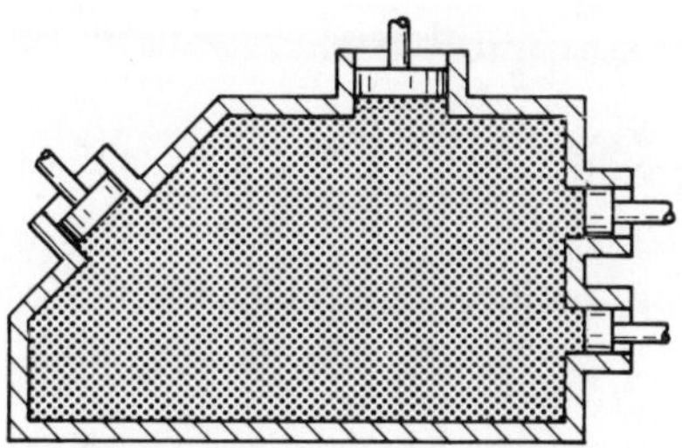

Figure 16.5 *Pressure in Closed Vessel*

The pressure of a liquid in a closed vessel is the same for all points in the liquid (neglecting the variation resulting from the depth of liquid in the vessel, which is usually very small compared with the pressure applied by external forces). The pressure produced in the liquid by any one of the pistons, is therefore, immediately transferred to all the other pistons.

ences between liquids and gases), the pressure is transmitted through the liquid to every other piston. These other pistons are now capable of overcoming resistances equal to the area of the piston × the pressure of the liquid. By suitably arranging the piston diameters, it is possible to use this principle of transmission of pressure in liquids as a basis for certain useful machines.

Fundamentally, two cylinders of large diameter A and small diameter a are fitted with pistons and connected together by a pipe, Fig. 16.6. The cylinders and the connecting pipe are full of liquid. A force P N is applied to the small piston producing a liquid pressure of $\frac{P}{a}$ N/m². This pressure

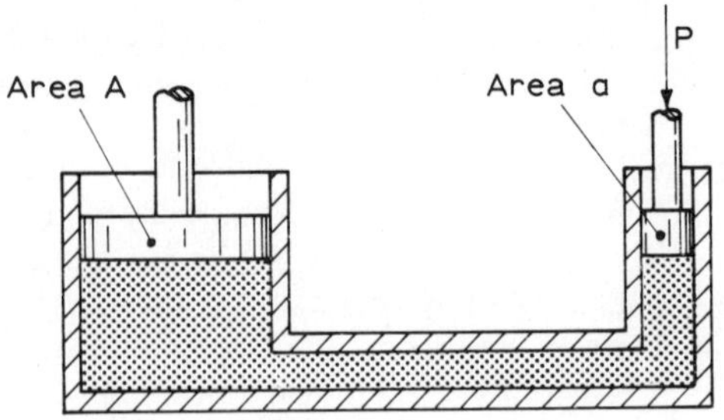

Figure 16.6 *Principle of the Hydraulic Machine*

A force P applied to the smaller piston produces a pressure in the liquid. This pressure is transferred to the larger piston, and can be used to overcome a big resistance.

is constant throughout the system, and produces a force on the large piston equal to pressure × area, or $\frac{P}{a} \times A$.

Hence, an effort P can overcome a resistance $P \times \frac{A}{a}$. This gives a mechanical advantage of $\frac{A}{a}$, which can be made any particular value by suitably choosing the piston diameters.

Hydraulic presses, hydraulic riveting machines, hydraulic brakes on cars, hydraulic jacks and quite a number of other familiar machines work on this principle.

In common with other machines, the distance travelled by the effort is much greater than the distance travelled by the load. The work done by the effort will be slightly greater than the work done in moving the load due to frictional forces. These forces are produced particularly by the special packings which are provided to prevent leakage of the liquid, under pressure, past the piston.

The manometer

The manometer is a very convenient instrument with which to measure pressure accurately. In its simplest form it consists of a straight open tube inserted into a horizontal pipe in which a liquid is flowing. The tube should project to the centre of the pipe, Fig. 16.7, and the immersed end of the tube should be 'square' with the direction of liquid flow in the pipe.

Liquid will rise in the tube to a height of h metres. As we have seen, this will indicate a pressure of $h\rho g$ N/m^2, where ρ is the density of the liquid in kg/m^3 and g is 9·81 m/s^2.

If the tube is inclined at an angle θ, a more accurate reading can be obtained. In this case the pressure in the pipe is $h\rho g$ as before, but h is measured vertically. The reading along the length of the tube will be

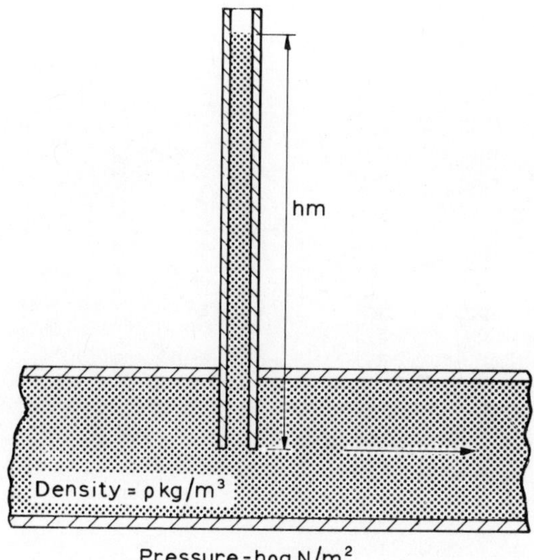

Figure 16.7 *Manometer*

$h_1 = \dfrac{h}{\sin \theta}$. If θ is 30°, h_1 will be equal to $2h$ and the gauge markings will be twice as far apart as they would be if the angle were 90°.

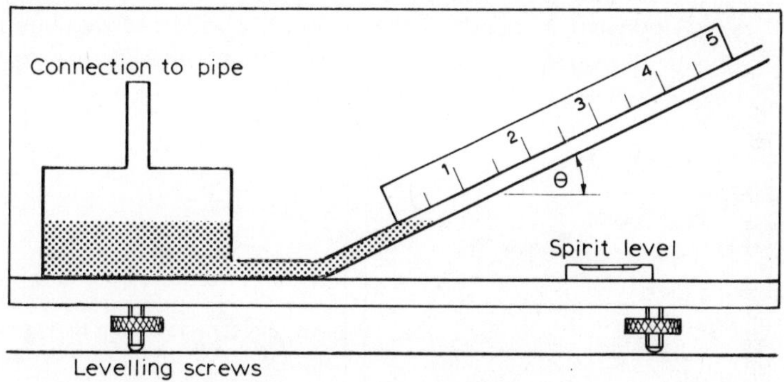

Figure 16.8 *Inclined Manometer*

An alternative form of manometer is the one shaped in the form of a U and filled with a suitable liquid. The height measured would be the difference in levels of the liquid in either arm of the U. The pressure in the pipe would be $h\rho g$ N/m^2, where ρ is the density of the liquid in the manometer. By choosing a suitable liquid (and therefore a suitable value of ρ) the value of $h\rho g$ can be varied to suit the pressure of the fluid in the pipe. This type of

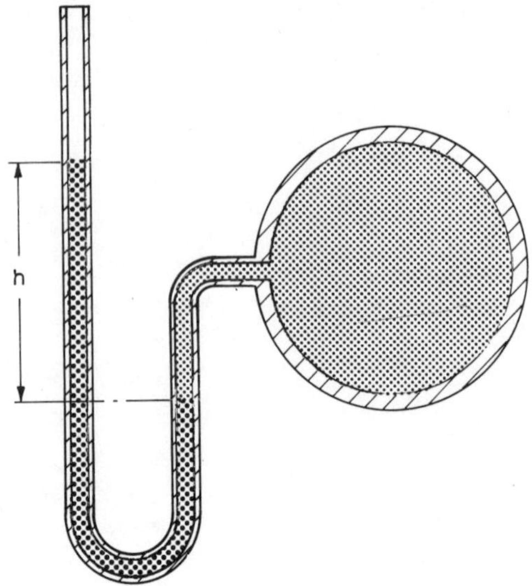

Figure 16.9 *U-tube Manometer*

manometer is useful for measuring the pressure of gases flowing in pipes. Remember that the actual, absolute pressure of the fluid in the pipe is equal to the pressure indicated by the manometer (gauge pressure) plus the pressure of the atmosphere.

If the pressure in the pipe is less than atmospheric pressure, then the level of the liquid in the left-hand arm of the manometer will be lower than that in the right-hand arm. In this case, the absolute pressure is equal to the atmospheric pressure minus the gauge pressure.

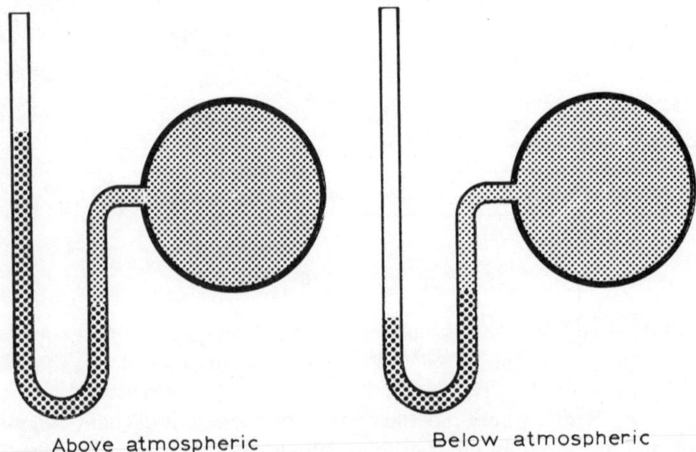

Figure 16.10 *U-tube Manometers*

Bellows-type pressure gauge

It is now possible to obtain a 'tube' of very thin walls made from phosphor bronze, but with the sides corrugated instead of being parallel, rather like a concertina. Both ends of the tube are sealed, with provision for the fluid to enter at one end which is fixed. Under pressure, the fluid causes the other end of the tube to move and this movement can be measured, usually by means of a pivoted lever and a calibrated scale. Alternatively, a corrugated diaphragm is used to cover the open end of a cylinder into which the fluid under pressure has access. Again the diaphragm moves, and the movement is indicated in a similar manner to that for the tube type.

The areas and thickness of the diaphragm and the length and thickness of the corrugated tube are the factors which determine the range of the gauge.

This type of gauge is suitable only for the smaller range of pressure measurement.

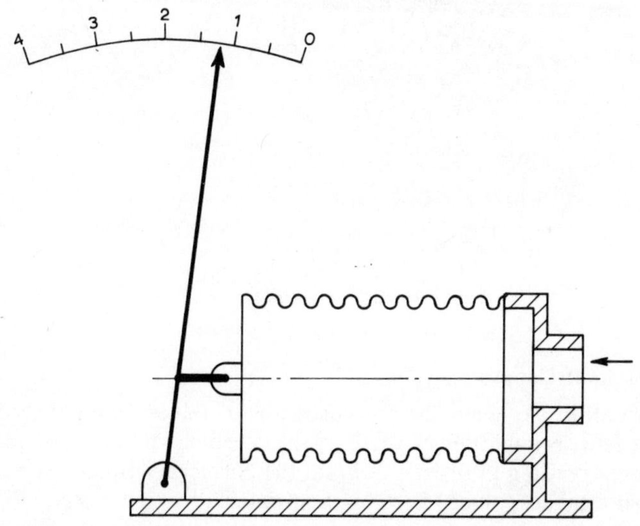

Figure 16.11 *Bellows Pressure Gauge*

Bourdon pressure gauge

When very high pressures are to be measured, it is necessary to use a more robust pressure gauge, the most common being the Bourdon type. This consists essentially of a flat tube of copper or bronze closed at one end and formed into the arc of a circle. The gas under pressure is connected to the open end of the tube which tends to straighten out. This movement is transferred through levers and pinions to a pointer which indicates the pressure on a scale graduated in suitable units. The physical dimensions of the tube, both cross-sectional area and tube thickness, are chosen to enable the gauge to be suitable for a given pressure range.

If the gauge is to be used for hot gases or steam, precautions must obviously be taken to prevent movement of the tube resulting from thermal expansion.

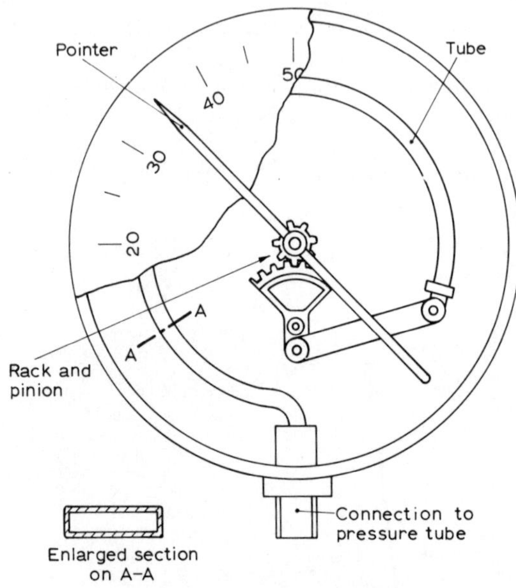

Figure 16.12 *Bourdon Pressure Gauge*

Calibration of Gauges

By calibration we mean the establishment of the accuracy of the instrument; the determination of the correction which should be made to the instrument reading in order to obtain the correct reading.

Pressure gauges can be tested by means of dead-weight testers, Fig. 16.13. Weights are applied to a piston in an oil-filled cylinder. The gauge is mounted in such a way that it records the pressure in the oil. Thus a known pressure is produced in the oil and this pressure is compared with that recorded on the gauge.

Comparisons are made over a series of pressures and the observations recorded. Sometimes the error shown on the gauge varies erratically. In other cases, the error is a progressive one, varying uniformly over the range of pressures.

Typical results are shown in the table.

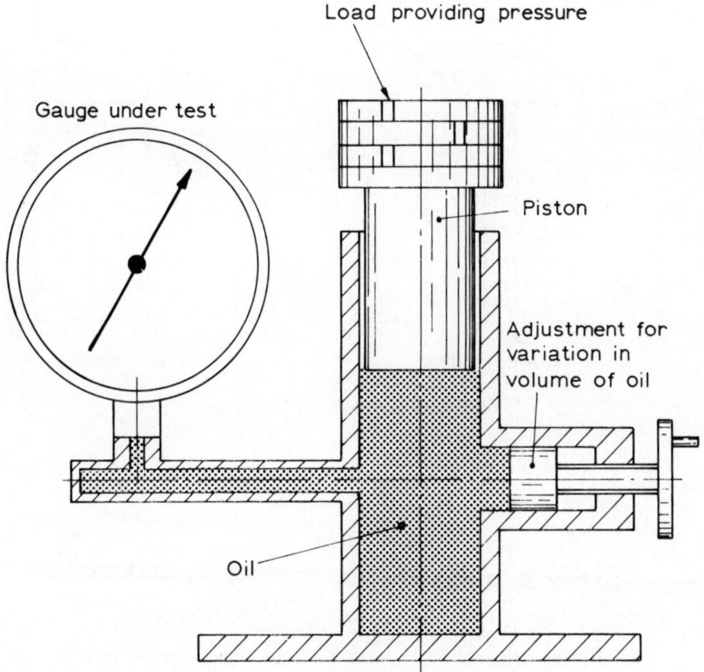

Figure 16.13 *Pressure Gauge Calibration Test*

actual pressure	0	20	40	60	80	100
gauge reading	0	22	40	58	79	100
error	0	+2	0	−2	−1	0
correction required	0	−2	0	+2	+1	0

The last row of figures, the correction required, indicates the addition or subtraction which has to be made to the gauge reading in order to obtain the correct reading. These results are often presented in the form of a graph which plots either the actual pressure against the gauge reading, or the correction against the gauge reading, Fig. 16.14.

Example 16.3

An hydraulic press has a small piston 50 mm in diameter and a large piston 200 mm in diameter. The small piston exerts a force of 2 kN during its travel of 150 mm.

Calculate:
(*a*) the resistance which the large piston can overcome,
(*b*) the distance travelled by the large piston,
(*c*) the velocity ratio, and
(*d*) the mechanical advantage of the press if its efficiency is 85 per cent.

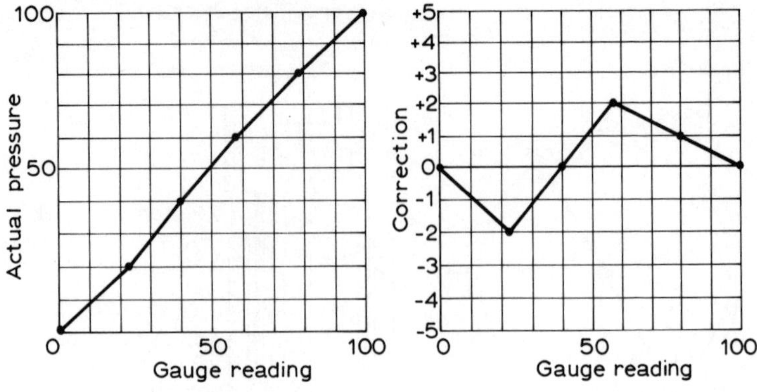

Figure 16.14 *Calibration Test Results*

$$\text{Pressure produced by small piston} = \frac{\text{force}}{\text{area}}$$

$$= \frac{2 \times 4}{\pi \times 0{\cdot}05^2}\ \text{kN/m}^2$$

$$= 1020\ \text{kN/m}^2.$$

(*a*) $$\text{Force transmitted to large piston} = \text{pressure} \times \text{area}$$

$$= 1020 \times \frac{\pi}{4} \times 0{\cdot}2^2\ \text{kN}$$

$$= 32\ \text{kN}.$$

An alternative solution to (*a*) would be:

$$\text{Force transmitted to large piston} = \text{force on small piston} \times \left(\frac{D}{d}\right)^2$$

where D is the diameter of the large piston and d is the diameter of the small piston.

(*b*) Volume of liquid moved by small piston is equal to volume displaced by large piston.

$$\therefore \quad \frac{\pi}{4} \times 0{\cdot}05^2 \times 0{\cdot}15 = \frac{\pi}{4} \times 0{\cdot}2^2 \times S$$

where S is the distance travelled by the large piston

whence $$S = 0{\cdot}009\,4\ \text{m}$$
$$= 9{\cdot}4\ \text{mm}.$$

(*c*) $$\text{Velocity ratio} = \frac{\text{distance moved by small piston in same time}}{\text{distance moved by large piston}}$$

$$= \frac{150}{9{\cdot}4}$$

$$= 16.$$

(*d*)
$$\frac{\text{Mechanical advantage}}{\text{Velocity ratio}} = \text{efficiency}$$
$$\frac{\text{MA}}{16} = 0{\cdot}85$$
$$\text{MA} = 16 \times 0{\cdot}85$$
$$= 13{\cdot}6.$$

Resistance overcome by the large piston is 32 kN, the distance travelled by large piston is 9·4 mm. The velocity ratio of the press is 16, and under the conditions given, the mechanical advantage is 13·6.

Example 16.4

The liquid in a manometer connected to a pipe is mercury, density $13{\cdot}6 \times 10^3$ kg/m^3. What is the absolute pressure in the pipe if the difference between the levels of the mercury is 500 mm? Atmospheric pressure is 1·01 bar.

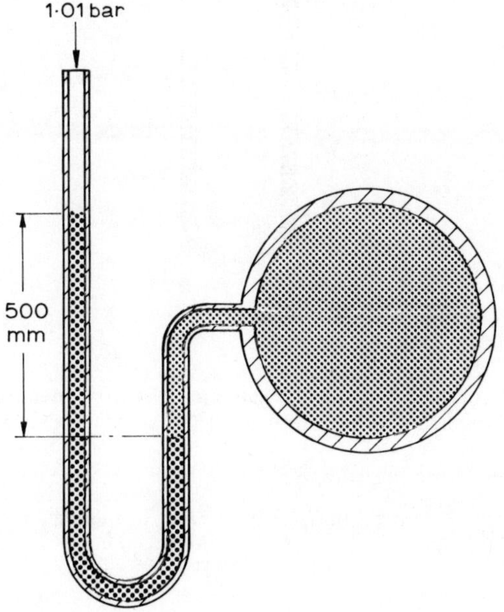

Figure 16.15

$$\text{Pressure due to head of liquid } h = \text{density} \times g \times \text{head}$$
$$= 13{\cdot}6 \times 10^3 g \times 0{\cdot}5$$
$$= 67\,000 \text{ N/m}^2$$
$$= 0{\cdot}67 \text{ bar.}$$

This is gauge pressure.

The absolute pressure in the pipe is equal to the gauge pressure + atmospheric pressure, and would equal 0·67 + 1·01 = 1·68 bar.

Absolute pressure in the pipe is 1·68 bar (168 kN/m^2).

Example 16.5

A U-tube contains mercury of density 13 600 kg/m². A quantity of alcohol of density 790 kg/m² is poured into one arm until the difference in the levels of mercury in the two arms is 38 mm, Fig. 16.16. What is the height of the column of alcohol?

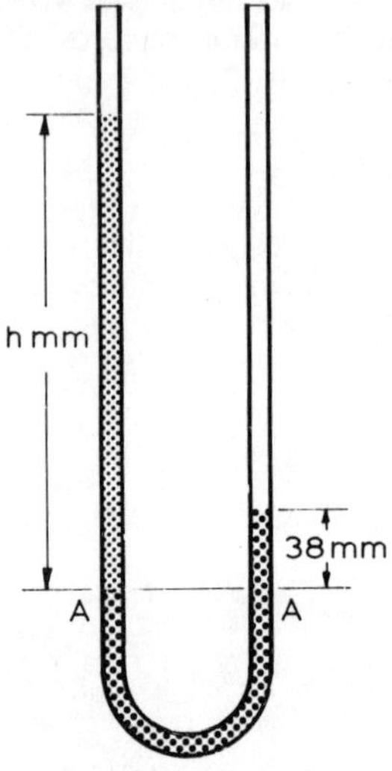

Figure 16.16

The pressure in each limb at the same level will be identical. Consider the pressure along the line AA, drawn through the point where the alcohol and the mercury meet.

Neglect atmospheric pressure, which will have the same effect on both arms.

Pressure in right-hand column due to 38 mm of mercury pressure

$$= 13\,600 \times 9{\cdot}81 \times \frac{38}{1000}$$

$$= 5065 \text{ N/m}^2.$$

Pressure in left-hand column due to h mm of alcohol pressure

$$= 790 \times 9{\cdot}81 \times \frac{h}{1000}$$

$$= 7{\cdot}75\, h \text{ N/m}^2.$$

These pressures are equal.

$$7{\cdot}75\, h = 5065$$
$$h = 654 \text{ mm}.$$

The height of the column of alcohol is 654 mm.

N.B.—You will see from the above equation that

$$h = 38 \times \frac{13\,600}{790} = 654 \text{ mm.}$$

Work done by pressure in cylinder

One important method of controlling a machine involves the use of hydraulic or pneumatic transmission of power. Oil or air or gas under pressure is injected into a cylinder containing a piston which is free to move. The piston is forced out of the cylinder and can be used to overcome a resistance.

If the pressure in the cylinder is p N/m^2 and the area of the cylinder and therefore of the piston is a m^2, the force exerted by the fluid on the piston is pa newtons, Fig. 16·17. If the piston moves l metres, the work done by the fluid on the piston is pal newton metres. This assumes that the pressure remains constant, but in many cases the pressure varies. As the piston moves, the volume occupied by the fluid changes and, in the case of a gas, this may cause the pressure to fall. In an internal combustion engine the pressure is produced by the explosion of the gas mixture in the cylinder. This explosion forces the piston out of the cylinder which in turn causes a drop in the gas pressure.

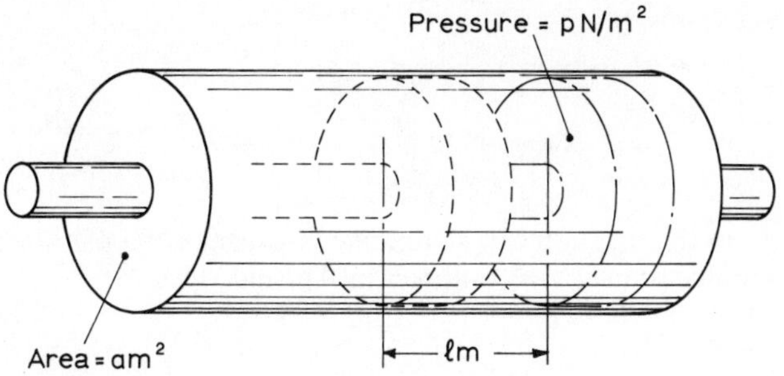

Figure 16.17 *Work done in Cylinder*

It is possible to obtain a visual indication of the variation of pressure in the cylinder. This can be done electronically by means of an oscilloscope operated by a pressure transducer, or it can be shown by an indicator attached to the cylinder. The cylinder is connected to a piston in the indicator. Any variation in pressure in the cylinder causes the indicator piston to move. Its movement is controlled by a suitable spring whose stiffness depends upon the pressure range under consideration. The piston movement is transferred through a system of levers to a stylus which draws a diagram on a piece of prepared paper mounted on a revolving drum. The movement of the paper corresponds on a reduced scale to the movement of the piston.

The vertical movement indicates the change of pressure and the horizontal movement corresponds to the movement of the piston. Fig. 16.18 shows a typical diagram taken from an engine. The area of the diagram divided by the base will give the mean height of the diagram. This is proportional to the mean or average pressure in the cylinder during the stroke. The indicator spring is rated as corresponding to a certain pressure per mm of compression. This enables the mean height of the diagram to be converted to mean pressure.

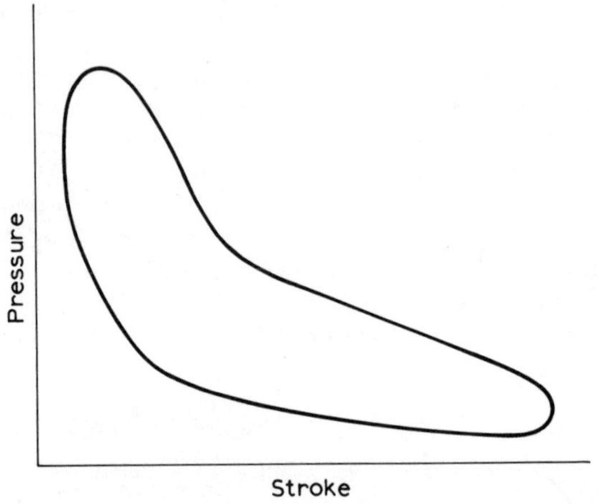

Figure 16.18 *Indicator Diagram*

The area of the diagram can be obtained by means of a planimeter or by suitable graphical means, as outlined in Appendix 1.

Example 16.6

The average pressure in the cylinder of an engine is 540 kN/m². If the diameter of the piston is 0·3 m and the stroke is 0·4 m, calculate the power of the engine if it runs at 240 rev/min and has only one working stroke each revolution

$$\text{Area of piston} = \frac{\pi}{4} \times 0{\cdot}3^2$$

$$= 0{\cdot}071 \text{ m}^2.$$

$$\text{Average force on piston} = \text{pressure} \times \text{area}$$

$$= 540 \times 0{\cdot}071$$

$$= 38{\cdot}4 \text{ kN}.$$

$$\text{Work done per stroke} = \text{force} \times \text{distance}$$

$$= 38{\cdot}4 \times 0{\cdot}4$$

$$= 15{\cdot}32 \text{ kNm}$$

$$= 15{\cdot}32 \text{ kJ}.$$

$$\text{Work done per second} = 15{\cdot}32 \times \frac{240}{60} \text{ kW}$$
$$= 61{\cdot}3 \text{ kW}.$$

Power of engine is 61·3 kilowatts.

Example 16.7

Fig. 16.19 shows an indicator diagram taken from an engine which has a cylinder 90 mm in diameter and a 100 mm stroke. The indicator spring strength corresponds to a pressure of 100 kN/m² per mm compression. The engine gives 600 explosions or cycles per minute. Calculate the power of the engine.

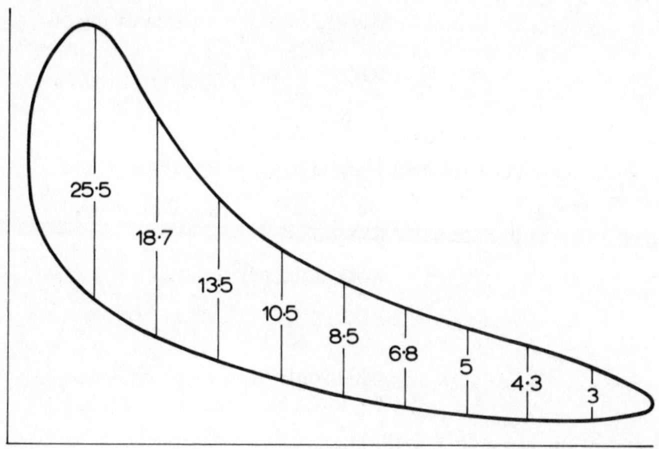

Figure 16.19

Using the method known as Simpson's Rule (Appendix 1) the base of the diagram is divided into 10 equal parts and the length of the ordinate at each division is measured. Ordinates are numbered from 1 to 11.

Number	1	2	3	4	5	6	7	8	9	10	11
Length mm	0	25·5	18·7	13·5	10·5	8·5	6·8	5	4·3	3	0

The width of the division can be taken as 1, which means that the length of the diagram will be 10.

$$\text{Area of diagram} = \frac{\text{width of one strip}}{3} \times \text{sum of} \begin{Bmatrix} \text{first} + \text{last ordinates} \\ 2 \times \text{sum of odd ordinates} \\ 4 \times \text{sum of even ordinates} \end{Bmatrix}$$

$$\text{First} + \text{last} = 0 + 0$$
$$= 0$$
$$2 \times \text{sum of odd ordinates} = 2(18{\cdot}7 + 10{\cdot}5 + 6{\cdot}8 + 4{\cdot}3)$$
$$= 80{\cdot}6$$
$$4 \times \text{sum of even ordinates} = 4(25{\cdot}5 + 13{\cdot}5 + 8{\cdot}5 + 5 + 3)$$
$$= 222{\cdot}0$$

$$\text{Area} = \tfrac{1}{3} \times (0 + 80{\cdot}6 + 222)$$
$$= \tfrac{1}{3} \times 302{\cdot}6$$
$$= 100{\cdot}9$$

$$\text{Average height of diagram} = \frac{\text{area}}{\text{base}}$$
$$= \frac{100{\cdot}9}{10}$$
$$= 10{\cdot}09 \text{ mm.}$$

$$\text{Average (mean) pressure} = 10{\cdot}09 \times 100$$
$$= 1009 \text{ kN/m}^2.$$

$$\text{Average force on piston} = \text{average pressure} \times \text{area of piston}$$
$$= 1009 \times \frac{\pi}{4} \times \left(\frac{90}{1000}\right)^2$$
$$= 6{\cdot}4 \text{ kN.}$$

$$\text{Work done per stroke} = \text{average force} \times \text{length of stroke}$$
$$= 6{\cdot}4 \times 1000 \times \tfrac{100}{1000} \text{ Nm or joules}$$
$$= 640 \text{ joules.}$$

$$\text{Power} = \text{work done per second}$$
$$= 640 \times \frac{600}{60}$$
$$= 6400 \text{ watts}$$
$$= 6{\cdot}4 \text{ kW.}$$

The power of the engine is 6·4 kilowatts.

EXERCISE 16

1 A tank of water is 50 m long, 20 m wide and 10 m deep. The depth of water in the tank is 8 m. Calculate:
(*a*) the pressure of the water on the base in kN/m^2,
(*b*) the total force on the base in kN.

2 In a hydraulic machine, a force of 200 N is applied to a piston 10 mm in diameter. The larger piston has a diameter of 200 mm. Calculate:
(*a*) the pressure in the liquid, and
(*b*) the resistance which the larger piston can overcome assuming that the machine is perfect.

3 A hydraulic accumulator with a ram 400 mm in diameter is connected with a hydraulic press whose ram is 800 mm in diameter. The load on the accumulator is 700 kN.
(*a*) What is the pressure, in kN/m^2, on the ram of the accumulator?
(*b*) What total force, in kN, would the press exert?
(*c*) Calculate the head equivalent to the pressure on the ram of the accumulator. [U.E.L.]

4 (*a*) What do you understand by the term 'head of water'?
(*b*) A hydraulic accumulator ram is 0·5 m in diameter. The dead load,

including the weight of the ram itself, is 270 kN. Calculate the pressure in kN/m^2 in the service mains.

(c) If this hydraulic pressure is used to work a hydraulic riveter, the plunger of which is 40 mm in diameter, calculate the effective force exerted on the rivet. [U.E.I.]

5 A water tank 30 m long and 10 m wide contains water to a depth of 4 m. What will be the pressure intensity on the base in kN/m^2?

6 The average pressure in the cylinder of an engine is 210 kN/m^2. The diameter of the piston is 0·25 m and the stroke is 0·35 m. Calculate the power of the engine if it makes 200 working strokes per minute.

7 An indicator diagram is divided into 10 equal strips and the sum of the mid-ordinates of the strips is 197 mm. The spring used in the indicator is such that a pressure of 10 kN/m^2 will deflect the spring by 1 mm. What is the average pressure in the cylinder?

If the engine makes 150 working strokes per minute, the piston diameter is 0·35 m and the stroke is 0·55 m long, determine the power of the engine.

8 A constant pressure of 450 kN/m^2 acts on a piston which is 0·3 m in diameter. If the stroke of the engine is 0·6 m, calculate the work done during the stroke.

9 An indicator diagram is divided into 10 equal strips and the height of the ordinate (in mm) at each division is as follows:

0, 21·2, 14·9, 10·4, 8·1, 6·7, 5·0, 4·1, 3·6, 2·6, 0.

What is the mean height of the diagram? If the indicator spring is rated at 120 kN/m^2 per mm compression, what was the mean pressure in the cylinder?

Calculate the power developed in the cylinder if its diameter is 80 mm, the stroke of the piston is 95 mm, and there are 500 cycles per minute.

10 Water is poured into a U-tube which contains mercury. The mercury rises

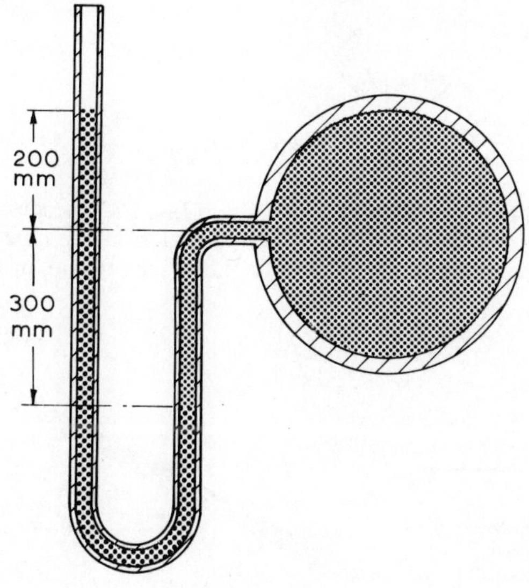

Figure 16.20

in one arm so that the difference in mercury levels is 80 mm. What is the height of the column of water?

11 A mercury U-tube manometer is used to measure the pressure of water in a pipe and is connected so that the water is in contact with the mercury, as shown in Fig. 16.20. The mercury in the right-hand limb is 300 mm below the centre line of the pipe and 200 mm above it in the left hand limb. What is the gauge pressure of the water in the pipe?

12 Details of mercury manometer readings are given in Fig. 16.21. Calculate the absolute pressure of the water in the pipe if the atmospheric pressure is 1 bar.

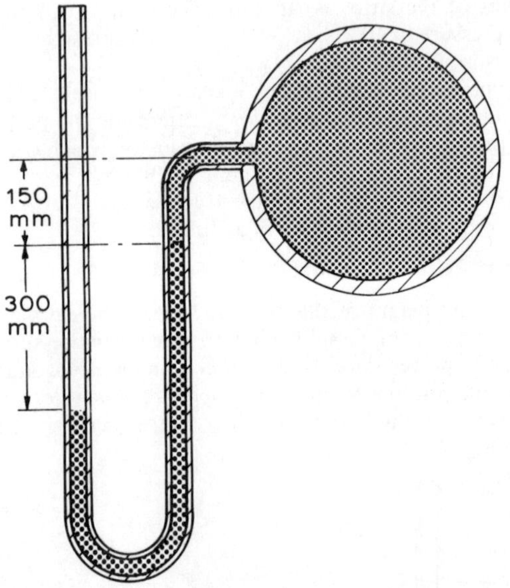

Figure 16.21

13 The inclined mercury manometer shown in Fig. 16.22 is connected to a pipe containing gas under pressure. If the inclination of the tube is 25° and the reading on the scale is 84 mm, what is the gauge pressure of the gas?

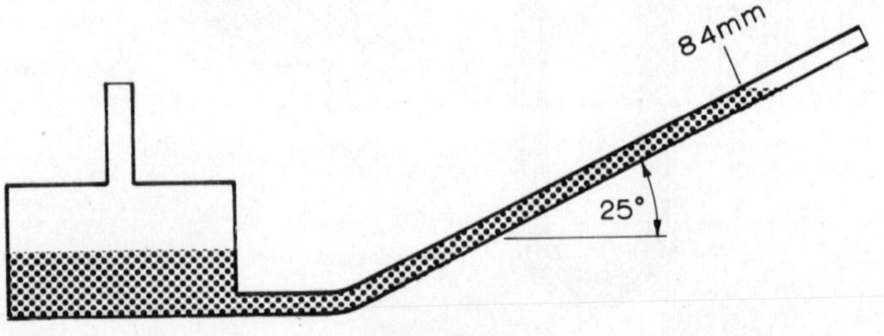

Figure 16.22

Chapter Seventeen
Measurement of Temperature and Transfer of Heat

In our studies so far we have considered questions such as engines pulling trains at given speeds over level tracks, cranes lifting weights, and boats moving at known speeds in known directions. We have given no thought to the way in which the power was to be produced to provide the conditions stated. The engine and crane require motors driven by electricity, produced at the power house by an alternator driven by a turbine. The boat requires its diesel engine with its supply of crude oil. In all these cases the motive power is supplied by an arrangement called a heat engine. The heat engine makes use of the heat energy which is released when coal is burned or when petrol vapour or oil vapour is ignited. It converts this energy into another form suitable for driving machinery. Before we can understand the manner in which these engines work, we must examine some of the fundamental ideas concerned with heat.

Effects of heat

Whenever heat is applied to a substance, certain effects are noticed. The following are the more important:

(1) *Change of dimension.* Liquids, gases and, more particularly, the metallic solids expand and contract when heat is added to or taken away from them.

(2) *Change of state.* Ice melts and becomes water which will boil and be converted into steam as heat is applied. The reverse sequence can be obtained if we remove heat in the steam. The steam will condense and the water which forms can then be frozen and ice produced.

(3) *Change of composition.* If heat is applied to grains of sugar they will turn into a brown caramel and ultimately into a black carbon. In this case however the process is not reversible: we cannot obtain sugar from carbon merely by the process of cooling it down.

(4) *Change of colour.* When an electric current is passed through a piece of tungsten wire, the wire becomes heated and if the current is large enough, the wire will glow red and ultimately white. When we switch on the electric light this all happens instantaneously. The reason why the wire gives out a bright light is because it changes its colour due to being heated by an electric current.

(5) *Electrical effect.* This is perhaps the least familiar of the effects caused by heat. If two dissimilar metals such as platinum and iridium are joined together and heat is applied to the junction, a small electric current is produced which could be measured by a sensitive galvanometer. The junction of the two metals is called a thermocouple.

Temperature

In our normal conversation the words 'hot' and 'cold' are sufficient to describe the state of a body from a heat point of view. Scientifically, however, we must be more explicit. We must be in a position to answer the question: how hot or how cold? We must have some method of stating the amount of hotness or coldness of a body. This is really what we mean by the temperature of a body.

DEFINITION. **Temperature is the degree of hotness or coldness of a body.**

In order to obtain an accurate idea of the temperature we must use some scientific instrument.

We shall obtain a much more accurate idea of the temperature if we observe some of the above effects of heat. For example, if we are out of doors on a winter's evening and we see a layer of ice on the ponds, we know that the temperature must be very low because ice only forms when the temperature reaches a low level. We cannot be certain of the exact temperature but we do know that it must be below a certain value. If we saw water poured on to an iron plate and the water was immediately converted into steam, then we should know that the plate was hot and that its temperature was above a certain value, although how much above we should have no means of knowing. In these two examples we are using the effect of heat which produces a change of state to give us an indication of the temperature. In the same way we could use any of the effects of heat. The blacksmith makes a lot of use of the colour effect. When he is tempering his tools he does not use any particular instrument to tell him when to plunge the tool into the water or the oil. He heats the tool until it glows a certain colour and he knows that the temperature has then reached the required value for successful tempering. Turning tools and chisels are normally tempered by heating them at the cutting end until the metal glows cherry-red and then quenching them in oil.

The most accurate method of measuring temperature is by making use of either the first or the last of the above effects of heat. The thermometer is an instrument for measuring temperatures by means of expansion of either liquids or metals. The pyrometer indicates temperature readings by means of the small electrical current produced when heat is applied to the thermocouple.

The thermometer

This consists of a glass tube having a very fine uniform bore. This is called a capillary tube. One end of the tube is sealed and formed into a bulb. A liquid, usually mercury, is inserted into the tube, the air is forced out and the other end of the tube sealed. The two most common liquids used in thermometers are mercury and alcohol. The choice depends upon the type of work which the thermometer has to do. For example, alcohol boils at a lower temperature than does water. Alcohol would therefore be quite unsuitable for thermometers required to measure temperatures as high as boiling water. Again, the mean temperature at the North Pole in January is considerably lower than the freezing point of mercury. Hence, if a mercury thermometer were taken to the North Pole at this time of the year the mercury would freeze solid and the thermometer would be no use for temperature measurements. Both these difficulties are overcome if a mercury thermometer is used for temperatures normally encountered in this country, and an alcohol thermometer is used on Arctic expeditions. Because alcohol is a colourless liquid, it is necessary to add some colouring, usually red, so that the level of the liquid in the tube can be read easily.

The pyrometer

The pyrometer is an instrument which measures the temperature of a body by measuring the small electrical current produced when heat is applied to a thermocouple. Fig. 17.1 shows a sectional arrangement of one type of pyrometer. Two wires of dissimilar metals are welded together at one

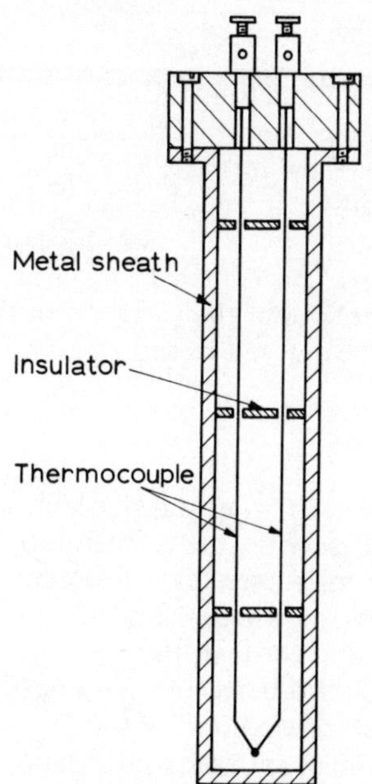

Figure 17.1 *The Pyrometer*
An instrument for measuring high temperatures by electrical means.

end and joined to separate terminals at the other ends. They are insulated from each other by means of mica washers and in some cases by passing through narrow bore porcelain tubes. Various pairs of metals are used, depending upon the temperature range required. Copper and constantan and nickel and nickel-chrome are two pairs frequently used. The delicate wires are mounted in a steel sheath for protective purposes. Specially designed connecting wires connect the pyrometer with an electrical galvanometer which measures the quantity of electrical current flowing in the circuit. The galvanometer dial is graduated directly in degrees.

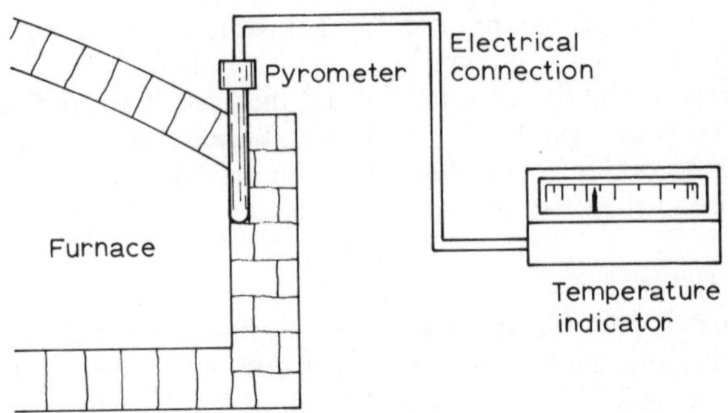

Figure 17.2

Diagram showing the arrangement of the pyrometer and connections for measuring the temperature of a furnace.

Fig. 17.2 shows how the pyrometer is used in measuring the temperature of an electrical furnace. The instrument is mounted in the side of the furnace to avoid reducing the available space of the furnace. The galvanometer can be mounted on an adjoining wall or it may be placed in a control office together with other instruments to allow for a central control of the heat treatment operations. A special recording gear is often used in conjunction with the pyrometer so that the variation of the temperature from hour to hour throughout the day can be shown graphically.

Optical pyrometer

An optical pyrometer can be used to measure temperatures which are high enough for visible light to be radiated, such as the temperature of molten metal in furnaces. In principle, the instrument is a telescope in which there is a small lamp, the brightness of whose filament can be varied by altering the electric current. Looking through the telescope at the heat source, the filament of the lamp will also be visible. By adjusting the voltage until the 'heat colour' or 'temperature colour' of the filament coincides with that of the heat source, the filament seems to disappear. The instrument is suitably calibrated so that the setting of the control potentiometer at which the filament 'disappears' gives a direct reading of the temperature. Alternatively, a milliammeter reading of the current flowing in the lamp circuit will give the required temperature.

Thermometer scales

The numbering of the divisions on a thermometer scale is a very arbitrary

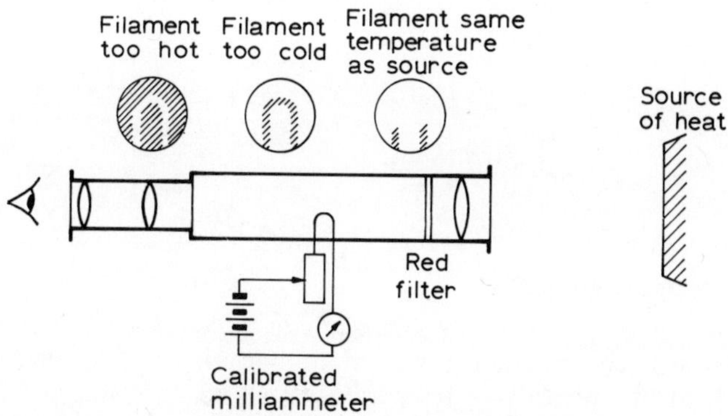

Figure 17.3 *Optical Pyrometer*

matter, based on two 'fixed points' at which a specific and easily reproducible temperature is defined as being so many degrees. For instance, the Celsius temperature scale is based on a lower fixed point, defined as 0°C, at which ice melts under normal atmospheric pressure, and an upper fixed point, defined as 100°C, at which water boils, again under normal atmospheric pressure. These two points 100 degrees apart effectively define the magnitude of a single degree on the Celsius scale, and any temperature above, below, or between the two fixed points can be defined in terms of this degree.

The earlier Fahrenheit scale defined 0°F as the temperature of a 'freezing-mixture' of ice and salt, which produced the lowest temperature obtainable in the laboratory in 1712, when Fahrenheit drew up his scale, and 100°F as the 'normal' temperature of the human body. On this scale, water freezes at 32°F and boils at 212°F. Unfortunately the temperatures of freezing-mixtures and of the human body are not fixed quantities so the scale was later redefined with 32°F and 212°F as the fixed points. On this revised scale the temperature we normally think of as 'body temperature' is 96·8°F.

In view of the uses made of the Fahrenheit scale, it might be of value to summarise the method of converting a Fahrenheit temperature to a Celsius temperature.

If °C = temperature in degrees Celsius
and °F = temperature in degrees Fahrenheit

then
$$°C = \frac{5}{9}(°F - 32)$$

Example

Convert (*a*) 131°F and (*b*) −310°F into °C.

(*a*)
$$°C = \frac{5}{9}(°F - 32)$$

$$= \frac{5}{9}(131 - 32)$$

$$\therefore \quad °C = 55 \qquad \text{i.e. } 131°F = 55°C$$

(*b*)

$$°C = \frac{5}{9}(-310 - 32)$$

$$\therefore \quad °C = -190 \qquad \text{i.e. } -310°F = -190°C$$

Absolute temperature scale

This dependence upon arbitrarily fixed points, which in turn are concerned with arbitrarily chosen liquids, is anything but scientific. A more rational approach to the question of temperature measurement was put forward by Lord Kelvin, who proposed that a zero of temperature should be found which would be independent of these physical properties. This basic zero of the temperature scale is known as the *absolute zero* and the scale based on it as the *thermodynamic temperature scale.*

Experimental work on the theory of gas laws has suggested that it is not possible to go below a temperature of −273°C (−273·15°C actually). The Kelvin scale of temperature starts at −273°C which it defines as 0 kelvin (0 K). The intervals on the Kelvin scale are the same as those on the Celsius scale, so that:

$$0\text{K} = -273°\text{C}$$
$$273\text{K} = 0°\text{C}$$
$$373\text{K} = 100°\text{C}$$

$$\text{Temperature K} = \text{temperature °C} + 273.$$

Temperatures on the Kelvin scale are often referred to as *absolute* temperatures. More work on these scales will be done in Chapter 18, Heat and Gases.

Note that the interval on the Kelvin scale is known simply as a kelvin and its symbol is K. The term 'degree kelvin' and the symbol '°K' are *not* used.

Accuracy and calibration

There are six accurately known temperatures which are used as 'fixed points' in the calibration of temperature measuring instruments, together with some fourteen secondary reference points. The six points are:

−189·2°C boiling point of oxygen
0°C melting point of ice
100°C boiling point of water
444·6°C boiling point of sulphur
960°C melting point of silver
1063°C melting point of gold.

Sealed containers of these substances are available which can be heated by built-in heating coils. The thermometer under test is inserted into an appropriate well in the container and the recorded temperature of the chosen substance is noted and compared with the standard figure. In the case of melting points, it is usual to raise the temperature above that at which the substance is known to melt and then allow it to cool. It will be seen that the falling temperature remains at one value for a short time. This is the temperature at which the substance melts and the thermometer reading is compared with the known value.

Whilst errors in liquid-in-glass thermometers can arise from the physical dimensions of the thermometer, such as variation in the size of the capillary bore, the more usual source of error results from the various depths at which the thermometer is inserted. If the instrument is immersed in a liquid so that only the bulb is covered, it is likely that a slightly different reading will be obtained than if the most of the thermometer is immersed in the liquid.

The mechanical equivalent of heat

We will have noticed that whenever any machining is being done the tool quickly becomes hot. In fact, in many operations a quantity of coolant is poured over the tool and workpiece to keep the temperature reasonably low, so that the operation can be carried out efficiently. There are many examples of heat being produced when movement takes place between two materials. A shaft rotating in a bearing can produce sufficient heat, if not sufficiently lubricated, to cause it to expand and bind itself in the bearing, or cause 'seizure.' Even if this extreme does not occur, we shall find that a bearing becomes appreciably warmer after the shaft has been running some time. In a worm gear drive the threads of the steel worm are continually sliding over the tooth surface on the bronze wheel and a considerable amount of heat is produced. It is part of the duty of the designer to make sure that the heat generated is reduced to a minimum and that which is generated is removed as quickly as possible. During the flight of space rockets, their speeds become so high that passage through the atmosphere is a relatively difficult matter. So much heat is generated, due to friction between the skin of the capsule and the atmosphere, that the temperature rises to 600°C and the projectile would be visible at night. A special heat shield has to be devised, therefore, to protect the capsule from destruction. Projectiles moving through the universe often pass through the atmosphere surrounding the earth. In their case they become white hot and are often seen moving across the sky and are given the name 'shooting stars.' As they pass out of the atmosphere they cool down and become invisible to the naked eye.

Now the immediate question is, where does heat come from and how is it produced? In all cases we shall find that the heat produced is dependent upon two factors:

(1) The speed of movement between the two surfaces in contact, e.g. drill and workpiece, shaft and bearing, worm and wheel, rocket and air.

(2) The resistance which one surface offers to the movement of the other surface.

The association of movement with resistance brings us back to the idea of work done. The work which the shaft has to do in overcoming friction in the bearing is converted into heat. The work done by rockets in overcoming the frictional resistance of the air is converted into heat. What is the relationship between the amount of work done and the quantity of heat produced?

Count Rumford, engaged on boring out cannon at the German Naval Arsenal, noticed that the metal removed by the boring tool was very hot. He arranged to carry out the boring operation under water using a very blunt tool. The heat produced was sufficient to boil the water in a very short time.

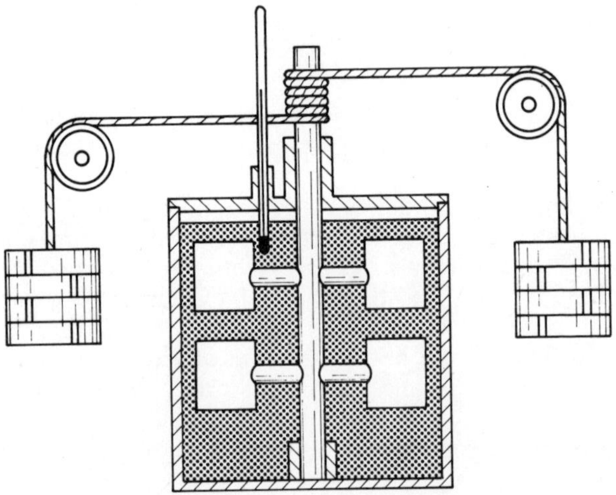

Figure 17.4 *Determination of Mechanical Equivalent of Heat*

The falling weights drive the paddles in the water. Due to the resistance of the water heat is generated and the temperature of the water rises. If the weight of the water is known together with the water equivalent of the calorimeter and the temperature rise it is possible to calculate the amount of heat produced by the falling weights.

In 1843, Joule performed experiments in Manchester to determine the relationship between the heat generated and the work done. The apparatus which he used is shown diagrammatically in Fig. 17.4. A paddle wheel immersed in water was rotated by weights. The weights were allowed to fall through a known distance, wound back to their initial position and then allowed to fall again. This was repeated a number of times. An arrangement was provided to allow the weights to be lifted without turning the paddles back. The amount of work done by the falling weights was found by multiplying the magnitude of the weights by the total distance through which they fell. The rise in temperature of the water was noted and the amount of heat produced by the paddles was calculated by multiplying the temperature rise by the mass of the water. Allowance was made for heat absorbed by the calorimeter and corrections were also applied to take into account radiation losses. More recent experiments have shown that 4187 joules (or newton metres) are required to raise the temperature of 1 kilogram of water by one degree Celsius.

Transfer of heat

It is very difficult for a body containing a quantity of heat to retain that heat unless it is suitably insulated. Insulation is provided either to prevent a body losing its heat to its surroundings, or to prevent a cold body from being raised to the temperature of its surroundings. The thermos flask prevents a hot liquid losing its heat or it prevents a cold liquid from receiving heat from an outside source.

Heat is transferred in one of three ways: conduction, convection or radiation.

Conduction

This is generally associated with solid bodies. Heat passes from one molecule to another within the solid *without the molecules moving*, rather like members of a human chain will pass buckets of water from one to the other from a water supply to a fire. There will probably be some increased vibration of the molecules but no general movement of them within the solid. Members of the chain move slightly backwards and forwards as the buckets are passed but they make no general movement along the line of the chain. So heat is passed along a poker inserted in a fire from the point to the handle. You can see the action of this type of heat transfer by using a rod of metal to which small solids have been fastened along its length by means of wax. One end of the rod is inserted into a source of heat. As the heat is conducted along the rod the wax is

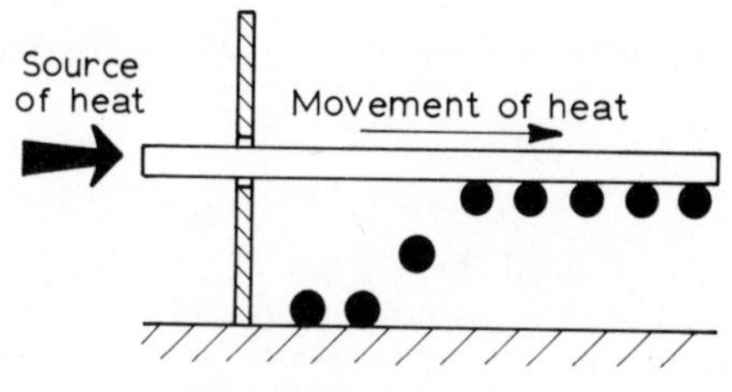

Figure 17.5 *Conduction*

melted and the solids fall off, Fig. 17.5. The first to fall is the one nearest to the supply of heat. The last to fall is the one furthest away from the heat.

The quantity of heat which will pass through a metal plate by conduction in a given time (t) can be shown experimentally to be dependent upon the following conditions:

1 The area of the plate (A). The greater the area the more heat can pass through.
2 The difference of temperature ($\theta_1 - \theta_2$) between the two sides of the plate
3 The thickness of the plate (x). More heat will pass through a thin plate than a thick one under similar conditions.
4 The duration of time under consideration (t). More heat will pass in a longer time than a shorter time.
5 The conductive characteristic of the material is represented by a coefficient k, the thermal conductivity of the material.

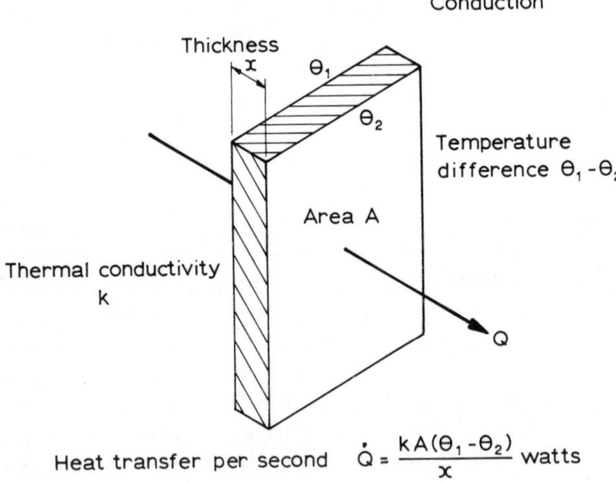

Figure 17.6 *Conduction of Heat*

The quantities are connected together by the equation.

$$\dot{Q} = \frac{kA(\theta_1 - \theta_2)}{x}$$

$\dot{Q}$ = quantity of heat flowing per second in watts. (N.B. The dot placed over Q indicates that this is a rate. $\dot{Q}$ (pronounced Q dot) means heat flow per second).
A = area of plate in square metres.
$\theta_1 - \theta_2$ = temperature difference in kelvin.
x = thickness of plate in metres.
t = time in seconds.
k = thermal conductivity of the material expressed in watts per metre per kelvin.

Heat flowing in t seconds is given by

$$Q = \dot{Q}t$$

$$= \frac{kA(\theta_1 - \theta_2)t}{x} \text{ joules.}$$

Thermal conductivities (W/m K)	
aluminium	209
brass	109
copper	385
cork	0·046
glass	1·05
iron	48·5
silver	419
steel	48
water	0·59

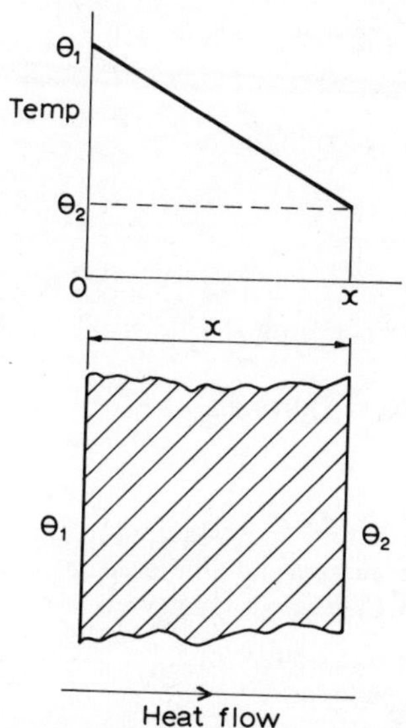

Figure 17.7 *Temperature Gradient*

Example 17.1

A metal plate is 125 mm thick and has a thermal conductivity of 60 W/m K. Calculate the quantity of heat flowing per hour through the plate if the temperature on one side is 10 K higher than on the other side. Express the result per square metre area of plate.

$$\text{Rate of flow of heat } \dot{Q} = \frac{kA(\theta_1 - \theta_2)}{x} \text{ joules per second}$$

$$= \frac{60 \times 1 \times 10}{125 \times 10^{-3}} \text{ joules per second per m}^2$$

$$= 4800 \text{ joules per second per m}^2.$$

$$\text{Quantity flowing per hour} = 4800 \times 60 \times 60$$

$$= 17\,280 \text{ kJ per square metre.}$$

Quantity of heat flow is 17 280 kJ/m² per hour.

Example 17.2

A glass window is 3 m high and 1·5 m wide and has a thickness of 8 mm. What quantity of heat will flow through the glass per hour if the temperatures of the glass are 2°C on the outside and 10°C on the room side. The thermal conductivity of glass is 1·05 W/mK.

$$\text{Area of window } A = 3 \times 1{\cdot}5$$

$$= 4{\cdot}5 \text{ m}^2$$

$$\text{Temperature difference} = (10 - 2)^\circ\text{C}$$

$$= 8 \text{ K}$$

$$\text{Heat flow per second } \dot{Q} = \frac{kA(\theta_1 - \theta_2)}{x}$$

$$= \frac{1{\cdot}05 \times 4{\cdot}5 \times 8}{8 \times 10^{-3}}$$

$$= 4725 \text{ joules per second}$$

$$\text{Heat flow per hour} = 4725 \times 60 \times 60$$

$$= 17\,010 \text{ kJ.}$$

Heat flow through the glass is 17 010 kJ per hour.

Example 17.3

Determine the area of plate 20 mm thick required to allow heat to be transferred at the rate of 1000 kW if the temperature difference anticipated between the inner and outer plates is to be 15°C.

Take the value of k for steel as 50 W/m K.
Let A be the area of the plate in m²

then
$$\dot{Q} = \frac{kA(\theta_1 - \theta_2)}{x} \text{ watts}$$

$$1\,000\,000 = \frac{50 \times 15A}{20} \times 1000$$

$$\therefore \quad A = 26{\cdot}7 \text{ m}^2.$$

Area of plate required is 26·7 m².

Example 17.4

A plate of aluminium 600 mm square and 5 mm thick is fastened to a sheet of brass 600 mm square and 10 mm thick. Assuming a good thermal contact between the two, determine the quantity of heat conducted through the plates if the face of the aluminium sheet is at a temperature of 150°C and that of the brass sheet is at 80°C. What will be the temperature at the joint face?

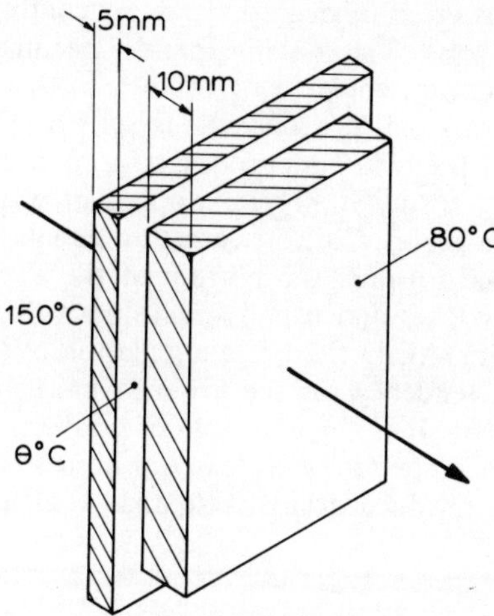

Figure 17.8

Thermal conductivity of aluminium is 210 W/m K.
Thermal conductivity of brass is 100 W/m K.
Let the temperature of the interface be θ°C.

Heat flow through aluminium will equal that through brass, i.e. $\dfrac{kA(\theta_1 - \theta_2)}{x}$ will be the same for each plate.

$$\text{Area of sheet} = 0{\cdot}6 \times 0{\cdot}6 = 0{\cdot}36 \text{ m}^2$$

$$\text{Rate of flow through aluminium} = \frac{210 \times 0{\cdot}36 \times (150 - \theta) \times 1000}{5}$$

$$\text{Rate of flow through brass} = \frac{100 \times 0{\cdot}36 \times (\theta - 80) \times 1000}{10}$$

$$\therefore \quad \frac{210 \times (150 - \theta)}{5} = \frac{100 \times (\theta - 80)}{10}$$

$$\theta = 136{\cdot}5\text{°C}.$$

Substituting into the expression for brass (or aluminium),

$$\dot{Q} = \frac{100 \times 0{\cdot}36 \times (136{\cdot}5 - 80) \times 1000}{10}$$

$$= 204\,000 \text{ watts.}$$

Temperature at the interface will be 136·5° C and the heat flow per second will be 204 kilowatts.

Convection

In this form of heat transmission the molecules themselves move. Convection applies in general to liquids and gases. It occurs quite automatically. If a beaker of water is placed over a bunsen flame, the heat from the flame is transferred through the glass by conduction to the water in contact with the glass. The water expands, becomes less dense and is displaced by the heavier, colder water above. In this way a movement of the water, sometimes called a current, is set up. The warmed water moves to the top and the cold water to the bottom, in turn to be warmed and sent to the top. A similar process applies in warming a room by means of a convection heater. Cold air comes into contact with the heating element usually placed towards the bottom of the appliance. The air is warmed, expands and rises and usually continues to rise until it reaches the ceiling of the room. Again there is a circulation of air throughout the room, but there is a tendency for the lower layers of the room, at foot level, to be rather cold. It is the principle of convection currents which enables the hot water system to operate in our homes. The same principle enables chimneys to produce draughts so that coal and coke fires can operate efficiently.

Radiation

Most of the heat which is lost by a body at a higher temperature than that of its surroundings is lost by radiation. Waves, which are considered to be electromagnetic, radiate from the heat source at the speed of light, 300 000 kilometres per second. When these waves strike other bodies some are reflected and some are absorbed by the body, whose temperature increases. This has no effect on the intervening substance, whose temperature is not affected. The heat we receive from the sun travels through space and reaches the earth by means of radiation. The space through which it passes has extremely low temperatures.

Materials which absorb heat easily are found to radiate heat easily and, conversely, material which does not absorb heat easily does not radiate heat easily. Generally, a substance which absorbs light well, i.e. a dark object, absorbs heat well. Similarly, a substance which reflects light well reflects heat well. Bright clothing which reflects light does not absorb well and consequently does not radiate heat well. Grey cast iron absorbs heat easily and likewise radiates heat easily. This means that a sphere of dull cast iron will loose heat by radiation much more quickly than a similar sphere of brightly polished copper containing the same quantity of heat.

A body which will absorb all the heat energy falling upon it is known as a black body. Such a body is the standard ideal radiator.

Stefan's Law

Stefan, an Austrian scientist, carried out a number of investigations into heat absorption and radiation in order to determine the law by which a body was cooled by radiation. Stefan's law (sometimes known as the Stefan-Boltzmann law) states that the heat radiated per second per unit area by a body is proportional to the fourth power of its absolute temperature. This is expressed as

$$E \propto AT^4 \text{ joules per second (or watts)}$$

where A is the surface area of the body, and
T is the absolute temperature of the body.

Alternatively $E = \sigma AT^4$ joules per second

where σ is the Stefan constant which has a value of $5{\cdot}67 \times 10^{-8}$ W/m^2 K^4 for the ideal black body.

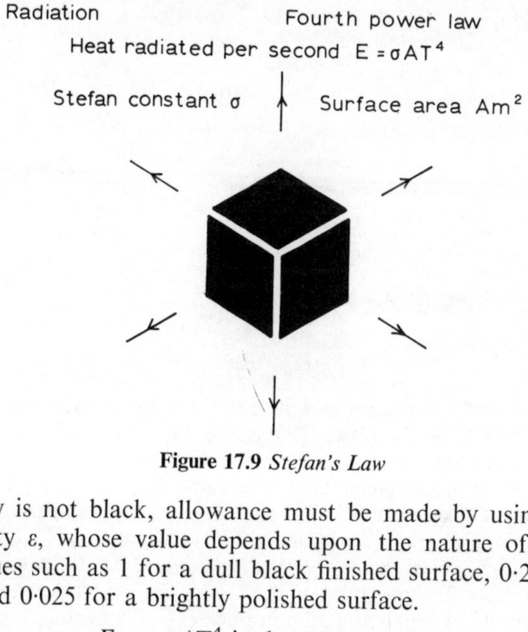

Figure 17.9 *Stefan's Law*

If the body is not black, allowance must be made by using a constant of emissivity ε, whose value depends upon the nature of the surface, having values such as 1 for a dull black finished surface, 0·28 for a lamp filament and 0·025 for a brightly polished surface.

Hence $E = \varepsilon\sigma AT^4$ joules per second.

Again, the condition is considered to be ideal in that the black body is considered to be isolated in a space at zero absolute temperature. The effect of the surrounding temperature (known as the ambient temperature) must be taken into account. If the temperature of the radiating body is T_1 and the ambient temperature is T_2, then the equation becomes

$$E = \varepsilon\sigma A(T_1^4 - T_2^4) \text{ joules per second.}$$

Example 17.5

The filament of a lamp has a temperature of 2000 K when 60 watts of power are supplied to it. What power will be needed if the temperature is to be 3000 K?

$$E = \varepsilon \sigma A T^4$$

The power supplied is proportional to the (absolute temperature)^4. If the temperature is to be increased by 3/2, the power must be increased by $(3/2)^4 = 81/16$.

$\therefore$ **Power required is $60 \times \frac{81}{16} = 303{\cdot}75$ watts.**

Example 17.6

The filament of a lamp operating at 2000 K radiates energy at the rate of 250 kW/m^2. What is the constant of emissivity of the filament?

$$E = \varepsilon \sigma A T^4$$

Energy radiated, $E = 250\,000$ W/m^2

Stefan constant, $\sigma = 5{\cdot}67 \times 10^{-8}$ W/m^{2}K^4

Area $A = 1$ m^2

Temperature $\theta = 2 \times 10^3$K

$$25 \times 10^4 = \varepsilon \times 5{\cdot}67 \times 10^{-8} \times 1 \times (2 \times 10^3)^4$$

$$\varepsilon = \frac{25 \times 10^4}{5{\cdot}67 \times 10^{-8} \times 16 \times 10^{12}}$$

$$= \frac{25}{5{\cdot}67 \times 16}$$

$$= 0{\cdot}276.$$

i.e. **Constant of emissivity is 0·27 for filament.**

EXERCISE 17

1 A steel plate 100 mm square and 50 mm thick has a temperature of 0°C on one face and 40°C on the other. Determine the amount of heat flow through the plate per minute.

2 A refrigerator is constructed in a box form and has a total surface area of 4 m^2. The thickness of the walls is 60 mm. The temperature inside the refrigerator is 5°C and that of the room in which it is kept is 35°C. The thermal conductivity of the material of the box is 0·046 W/m K. Calculate the quantity of heat flowing through the walls of the refrigerator per minute.

3 A slab of wood 1 m^2 area and 25 mm thick is firmly fixed to a similar slab of plastic 1 m^2 area and 15 mm thick. The temperature difference between the two outer faces is 40°C. How much heat is transferred through the composite material per min if the thermal conductivity of the wood is 0·12 W/m K and that of the plastic is 0·052 W/m K?

4 In a control mechanism, a rod of silver is fastened to a rod of brass so that good thermal contact is maintained between the two surfaces. The free end of the silver rod, which is 150 mm long, has a temperature of 120°C, while that of the brass rod is 20°C. What will be the length of the brass rod if the temperature at the junction has to be 70°C? The rods are insulated to prevent loss of heat along their sides.

5 A hot body, which can be treated as a black body, has an area of 0·1 m^2 and a temperature of 400 K. What is the rate of radiation from the body in watts?

6 If the body in question 5 is enclosed in a surrounding temperature of 500 K, and the constant of emissivity is 0·75, what will be the rate of radiation in watts?

7 The wall of a furnace has an opening in it of 200 mm² in cross-sectional area. Heat energy is radiated through the opening at a rate of 200 joules per second. Assuming that the furnace can be treated as a black body, calculate its temperature.

8 The filament of a lamp 300 mm long and 0·01 mm in diameter has a temperature of 2500 K. If the constant of emissivity is 0·2, determine the power radiated.

Chapter Eighteen
Heat and Gases

Kinetic theory of gases.

Whilst solids and liquids have their shape and size fairly well defined, gases have the property of being able to occupy fully any container into which they are placed. Once in the container, the physical properties of pressure and temperature are established in accordance with some familiar physical laws, which are based on a generally accepted idea known as the **kinetic theory of gases** The theory suggests that the molecules of the gas are moving within the container at a velocity dependent upon the temperature of the gas. The higher the temperature, the faster the molecules move. You may recognise here an application of the principle of the conservation of energy. Energy in the form of heat is applied to the gas, increasing the velocity of the molecules and thus increasing their kinetic energy.

Collisions take place between the molecules themselves and between the molecules and the containing vessel. As the moving molecules strike the vessel containing them, they rebound away from the boundary wall in the same way that a ball will rebound from a wall at which it has been thrown. The direction of the molecule's velocity has been changed and so has its momentum. We have seen that forces are involved in changes of momentum. The forces in this case come from the reaction of the wall of the container to the impinging molecule. The average of all these forces produces the effect which we call pressure. If the force required is so great that the container cannot provide it, it will burst 'under pressure'.

The relationship between the volume occupied by the gas and the temperature and subsequent pressure is determined by certain laws. In stating these laws one has in mind a 'perfect gas'. In fact the kinetic theory of gases assumes the existence of the perfect gas. In this gas:

(i) all the molecules are identical,
(ii) there is no mutual attraction or repulsion between the molecules,
(iii) the molecules are perfectly elastic, from which it follows that the impacts will be instantaneous and no losses of kinetic energy will occur, and
(iv) the total volume occupied by the molecules will be negligible compared with the volume of the gas.

No gas satisfies these conditions, but the observed behaviour of known gases is found to be very close to that predicted by the established gas laws.

You will have done some previous work on the expansion of gases, of which the following is a summary. The volume of a given mass of gas can be changed by altering:

(i) the pressure and keeping the temperature constant,
(ii) the temperature and keeping the pressure constant, or
(iii) both pressure and temperature.

Change of volume at constant temperature (Boyle's Law)

If the volume is changed by altering the pressure and keeping the **temperature constant,** the change is known as **isothermal** (which simply means 'at one temperature'). The change in volume is covered by a law known as Boyle's Law, named after the scientist who first discovered it.

If p_1 and v_1 are the original pressure and volume and p_2 and v_2 are the new pressure and volume, then the law of expansion (or contraction) is:

$$p_1v_1 = p_2v_2 = \text{constant.}$$

Which means that if we double the pressure we halve the volume, or that if we reduce the pressure to one-tenth of the original pressure then the volume will increase to ten times its original value.

Change of volume at constant pressure (Charles's Law)

If the volume is changed by altering the temperature and keeping the pressure constant, then the expansion follows a law known as Charles's Law. This law suggests that the volume of the gas would be doubled if the temperature were doubled, provided that we measured the temperature on the **absolute scale.**

If T_1 and v_1 are the original temperature and volume and T_2 and v_2 are the new temperature and volume, and provided that T_1 and T_2 are the absolute temperatures, i.e. we have added 273 degrees to the normal Celsius reading, then the law of expansion is:

$$\frac{v_1}{T_1} = \frac{v_2}{T_2}$$

Change of pressure, temperature and volume

The combined law which deals with the expansion of a gas whose pressure and temperature are changing together is

$$\frac{p_1v_1}{T_1} = \frac{p_2v_2}{T_2}$$

p_1, v_1 and T_1 are the original pressure, volume and temperature.
p_2, v_2 and T_2 are the new pressure, volume and temperature.
Remember, always use absolute temperatures.

Absolute and gauge pressures

The reading given on most types of pressure gauges is not the actual pressure of the gas, but the difference between the pressure of the gas and the pressure of the atmosphere. The gauge reading is usually called the 'gauge pressure,' while the actual pressure is called the 'absolute pressure.' These two pressures are related by the formula:

absolute pressure = gauge pressure + pressure of atmosphere.

The pressure of the atmosphere, which varies from day to day and from place to place, is taken as $1{\cdot}01 \times 10^2 \text{kN/m}^2$ or 1·01 bar at sea-level.

Standard temperature and pressure (s.t.p.)

The size of a gas container is no indication of the quantity of gas present. If you have a normal petrol container full of petrol, you know that you have approximately 10 litres of petrol. But to see a cylinder of oxygen gives you no idea of how much oxygen is in the cylinder. In fact, however much you take out of the cylinder it will still remain full, since the gas will always fill the container, although this is done at the expense of the pressure. When you see the letters 'MT' on an oxygen cylinder, what is meant is that the pressure of the oxygen in the cylinder is so reduced as to be unsuitable for further use.

In view of this variation of quantity with pressure (and also with temperature), it is important to have some standard of comparison. This is done by stating the volume of a gas at s.t.p., i.e. at a temperature of 0°C and a pressure of 1·01 bar. Hence, whatever volume of gas we may appear to have at certain temperature and pressure we can only get a true idea of the actual quantity of gas if we state the volume which it would have if its temperature were 0°C and its pressure atmospheric. Conversion to s.t.p. is done in the usual way using the combined gas law.

The gas constant R

The combined law dealing with change of pressure, volume and temperature which is usually written as $\frac{p_1 v_1}{T_1} = \frac{p_2 v_2}{T_2}$ can be expressed in the form

$$\frac{pv}{T} = \text{constant}$$

since for each condition of the gas the value of $\frac{\text{pressure} \times \text{volume}}{\text{temperature}}$ must be the same.

The constant is given the symbol R and is known as the gas constant. The value of R is determined by taking the pressure in N/m^2 and the volume of one kg of the gas measured in cubic metres.

One kilogram of air at 1·01 bar ($1{\cdot}01 \times 10^5 \text{ N/m}^2$) and a temperature of 0°C (273 K) occupies $0{\cdot}776 \text{ m}^3$.

$$\frac{pv}{T} = R$$

$$R = \frac{1{\cdot}01 \times 100\,000 \times 0{\cdot}776}{273}$$

$$= 287 \text{ Nm/kg K}$$
$$= 0{\cdot}287 \text{ kJ/kg K.}$$

Let $v =$ volume occupied by 1 kg of the gas
$V =$ volume occupied by m kg of the gas under same conditions

$\therefore \quad V = mv$

now $pv = RT$ (since our value for R as 287 was based on 1 kg of gas)

multiply by m

$$pvm = mRT$$
$$\therefore \quad pV = mRT$$

In this form, which is known as the characteristic equation for the gas, it is possible to determine the mass of the gas for given conditions of pressure, volume and temperature. Alternatively, it is possible to determine the volume which will be occupied by a certain mass of gas at a known temperature and pressure.

Units of R

p is in N/m^2
v is in m^3
m is in kg
T is in K

$$R = \frac{pv}{mT} = \frac{N/m^2 \times m^3}{kg \times K}$$

$$= \frac{Nm}{kg \times K}$$

The units of R are Nm per kg per K,
or J per kg per K,
or J/kg K.

Mixtures of gases: Dalton's Law of partial pressures

A problem sometimes arises when gases, either alike or different, are in separate vessels A and B under independent conditions of volume and pressure and are allowed to mix by opening a connecting valve. Generally we require to know the final common pressure of the mixture occupying both vessels. The law governing the final conditions is Dalton's Law of Partial Pressures and is based upon two assumptions:

(i) that the gases will not have any chemical action upon each other but will form a mixture, itself a gas, and
(ii) that the temperature remains constant throughout.

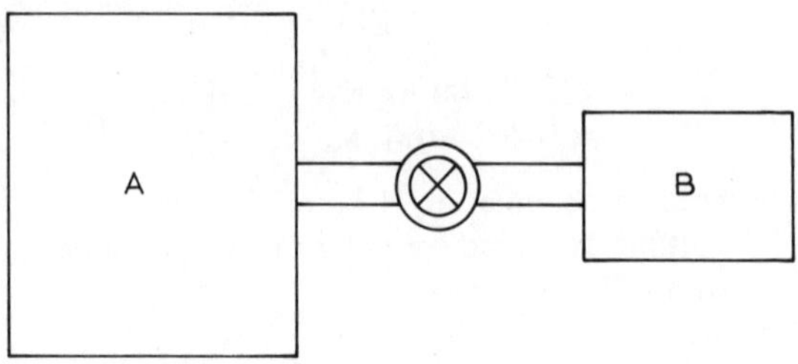

Figure 18.1

By partial pressure we mean the pressure which the gas would have if it were to occupy both vessels by itself. That is, assume vessel A contains a gas, while vessel B is completely empty. Open the valve and let the gas in A occupy both vessels A and B. Its pressure is then referred to as the partial pressure of gas in A. The partial pressure of the gas in B is similarly defined.

According to Dalton's law, the final common pressure of the mixture is the sum of the partial pressures of each gas, i.e. the sum of the pressures of each gas when that gas occupied both vessels alone.

There are many problems in the subject of applied heat which involve the mass of a given volume of gas. The determination of the mass of a gas is much more complicated than that of either a solid or a fluid. For example, you can measure the mass of a quantity of coal or petrol much easier than you can that of North Sea gas.

The principles used in calculating the mass of a known volume and temperature of a gas are (in addition to the laws of Boyle and Charles) Avogadro's hypothesis, relative molecular masses (or 'molecular weights') and the universal gas constant.

Elements and compounds

An element is a pure substance which cannot be chemically broken down into any simpler substance. There are 103 known elements, such as hydrogen, iron, gold and uranium. A compound is a chemical combination of two or more elements, for example water (hydrogen and oxygen) and carbon dioxide (carbon and oxygen).

Atoms and molecules

In searching into the nature of matter, man accepted for many years that it was made up of minute and indivisible particles known as atoms. In 400 BC, Democritus, a Greek philosopher, put forward a theory that all substances were ultimately built up from units he called 'atoms', a word which meant something that cannot be divided. Although the work of British physicists like Kelvin, J. J. Thompson, Rutherford and Chadwick has presented the structure of the atom in terms of electrons, protons and neutrons, the basis of nuclear physics, for our needs in thinking about simple physical and chemical problems it will be adequate for us to accept the concept of the atom as being the smallest division of an element having the properties special to it. The molecule is the smallest unit of an element or compound which exhibits the chemical properties of the substance *and has an independent existence.*

In some cases, the molecule contains only one atom; though it generally contains two or more.

Relative atomic and molecular mass

The mass of an atom or a molecule of an element is so very small that it would be impracticable to use it in calculations. It is, however, very important that we know the ratios of the masses of the atoms of the elements and of the molecules of the elements and their compounds. These ratios have been established and are known as relative atomic masses and relative molecular masses (or, to use the old terminology, 'atomic weights' and 'molecular weights').

The basis of the scheme of relative atomic masses is the atom of hydrogen which for our purpose is taken to have a mass of 1 unit. The relative atomic mass of oxygen is 16, which means that the oxygen atom is 16 times as heavy as the hydrogen atom.

A molecule of hydrogen contains two atoms so the relative molecular mass of hydrogen is 2. The molecule of oxygen also contains two atoms, so the relative molecular mass of oxygen is 32. The relative atomic mass of carbon is 12 and since a molecule of carbon contains only one atom, the relative molecular mass of carbon is also 12.

	Symbol	Relative atomic mass	Relative molecular mass
hydrogen	H	1	2
carbon	C	12	12
nitrogen	N	14	28
oxygen	O	16	32
sulphur	S	32	32

Avogadro's hypothesis

This theory, put forward by Avogadro in 1811, states that equal volumes of all gases measured at the same temperature and pressure contain the same number of molecules.

Consider the two vessels in Fig. 18.2. They have equal volumes, temperatures and pressures.

A contains hydrogen, whose relative molecular mass is 2, and

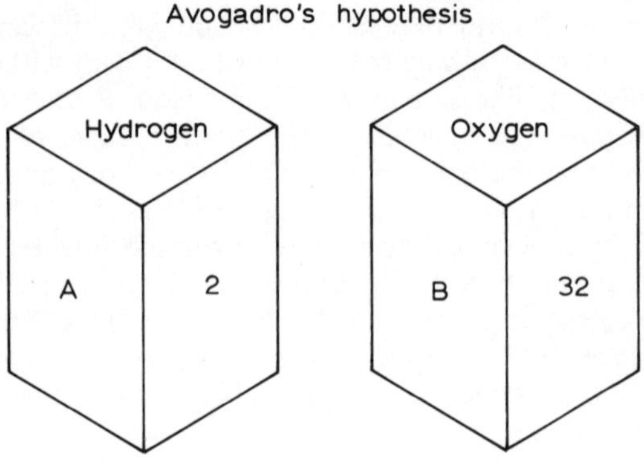

Figure 18.2 *Avogadro's Hypothesis*

B contains oxygen, relative molecular mass 32.

By Avogadro's law, both contain the same number of molecules, say N, since both volumes are equal.

Therefore the mass of A is $2N$ and the mass of B is $32N$.

It follows that the mass of equal volumes of different substances is proportional to the relative molecular masses of the substances.

A quantity of a substance whose mass in grams is numerically equal to the relative molecular mass of the substance is known as the mole (abbreviated to mol).

Similarly, the quantity whose mass in kilograms is numerically equal to the relative molecular mass of the substance is known as the kilomole (kmol).

If A contains 1 kmol of hydrogen, it will contain a mass of 2 kilograms.

If B contains 1 kmol of oxygen, it will contain a mass of 32 kilograms.

Since the volume of A and B are equal and they contain masses proportional to the relative molecular mass of the substance, it follows that A contains 1 kmol of hydrogen and B contains 1 kmol of oxygen.

Hence we have the fundamental relationship that 1 kmol of any gas will occupy the same volume under similar conditions of temperature and pressure.

Experiment shows that an average value for the volume of 1 kmol of gas at standard atmospheric pressure of 1·01 bar and 0°C is 22·4 cubic metres.

The universal gas constant, R_o

In the expression already established, $pv = mRT$, the use of the gas constant R, which has a different value for each gas, enables us to calculate the mass of that gas for given conditions of pressure, temperature and volume. The universal gas constant R_0 applies to all gases, and the determination of mass can be carried out provided we know the relative molecular mass of the gas under consideration.

Using the expression $pv = mRT$, let us consider the special case of a volume v_m occupied by 1 kmol of a gas which we know to be 22·4 m^3 when the pressure p is 1·01 bar and the temperature is 273 K. The mass of 1 kmol of any gas will be M kg where M is the relative molecular mass of the gas. Substituting these values in the above equation, we have:

$$pv_m = MRT$$

$$1{\cdot}01 \times 100\,000 \times 22{\cdot}4 = MR \times 273$$

$$MR = 8313 \text{ J/kmol K}$$

or

$$MR = 8{\cdot}31 \text{ kJ/kmol K}$$

This product MR is the universal gas constant R_0 and applies to all gases.

The gas constant R can be obtained from $R = R_0/M$ where M is the relative molecular mass of the gas.

For example, the relative molecular mass of hydrogen is 2, therefore the gas constant R for hydrogen is 8·31/2 or 4·15 kJ/kg K.

The relative molecular mass of air is accepted as 29 and the gas constant R for air will be 8·31/29 or 0·287 kJ/kg K.

Specific heat capacity, c

We shall define the specific heat capacity of a substance as the amount of heat required to raise the temperature of 1 kilogram of the substance by one kelvin. For example, the specific heat capacity of copper is 419 J/kg K. This means that 1 kg of copper requires an amount of heat equivalent to 419 joules in order to raise its temperature by one kelvin. The specific heat capacity of water is 4·19 kJ/kg K.

The same basic idea applies when we consider gases, except that consideration must be given to the fact that the gas can be heated under conditions of either constant volume or constant pressure.

Specific heat capacity of a gas at constant volume, c_v

Consider a unit mass of the gas in a closed vessel. Heat is supplied to the gas. The temperature rises and there is an increase in the velocity of the molecules of the gas, i.e. an increase in the kinetic energy of the gas. But no external work is done by the gas since it is not allowed to expand. Hence all the heat supplied is stored in the gas in the form of kinetic energy. The heat which is supplied to the unit mass of gas to raise its temperature by one degree is known as the specific heat capacity of the gas at constant volume and has the symbol c_v.

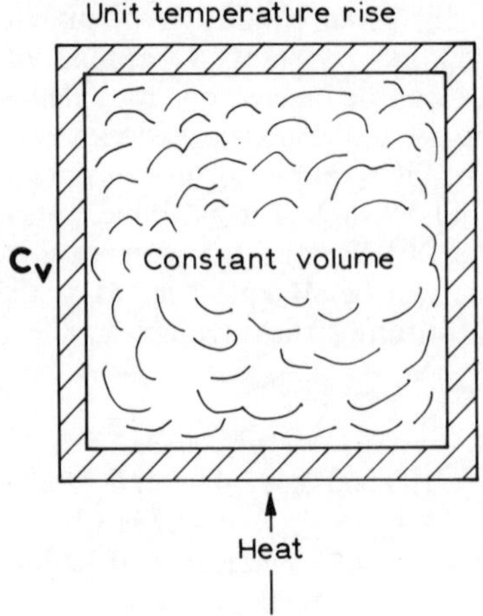

Figure 18.3 *Specific Heat Capacity at Constant Volume*

Specific heat capacity at constant pressure, c_p

Consider now the gas in a container fitted with a moveable piston on which a load is mounted. This load will provide a constant pressure in the gas. As heat is supplied, the temperature of the gas will increase as before. There will be an increase in the energy stored in the gas, but, in addition, the gas, now being free to expand, will move the loaded piston and therefore do work on the load.

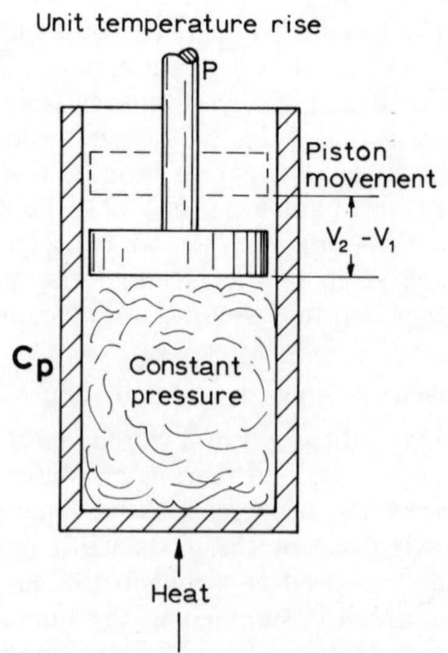

Figure 18.4 *Specific Heat Capacity at Constant Pressure*

Hence the heat supplied must:

1. raise the temperature at constant volume, and
2. do work in moving the piston.

For a temperature rise of one kelvin and a mass of 1 kg of gas, we can write:

$$c_p = c_v + \text{work done.}$$

Assuming that the area of the piston is 1 m² then the load P N on the piston will produce a pressure P N/m² in the cylinder. Also, if the volume of the gas increases from V_1 to V_2, it follows that the distance moved by the piston will be $(V_2 - V_1)$.

Work done in moving piston $= P(V_2 - V_1)$ joules.

Now $\quad PV_1 = mRT_1 = RT_1 \quad \text{since } m = 1$

also $\quad PV_2 = mRT_2 = RT_2$

$$= R(T_1 + 1).$$

$$\text{Work done} = P(V_2 - V_1)$$

$$= R(T_1 + 1) - RT_1$$

$$= R \text{ joules}$$

Substituting in the equation above we have

$$c_p = c_v + R$$

$$\text{or} \quad c_p - c_v = R$$

Specific heat capacity of gases kJ/kg K

	c_p	c_v
helium	5·19	3·11
hydrogen	14·22	10·06
oxygen	0·91	0·65
nitrogen	1·047	0·75
steam	2·02	
air	1·00	0·72
carbon dioxide	0·84	0·64
carbon monoxide	1·04	0·74

Example 18.1

If 20 m³ of gas at 50°C are heated at constant pressure until the volume is doubled, what will be the final temperature? [N.C.T.E.C.]

Since the pressure is constant, the gas will expand in accordance with Charles's Law.

$$\text{i.e.} \quad \frac{v_1}{T_1} = \frac{v_2}{T_2}$$

$v_1 = 20 \text{ m}^3 \qquad v_2 = 40 \text{ m}^3$

$T_1 = 50° + 273° \qquad T_2 = ?$

$= 323 \text{ K}$

$$
\begin{aligned}
T_2 &= \frac{v_2}{v_1} T_1 \\
&= \tfrac{40}{20} \times 323 \\
&= 646 \text{ K} \\
&= 373^\circ\text{C.}
\end{aligned}
$$

The final temperature of the gas will be 373° C.

Example 18.2

Two cylinders A and B contain quantities of a certain gas. The volume of A is 30 m³ and the pressure of the gas is 7 bar absolute at a temperature of 25°C. The volume of B is 35 m³ and the pressure of the gas in this cylinder is 5 bar absolute at a temperature of 20°C. Compare the quantities of gas in each cylinder.

Reduce the volume of the gas in each cylinder to s.t.p.

Cylinder A

$$
\begin{aligned}
p_1 &= 7 \text{ bar} & p_2 &= 1{\cdot}01 \text{ bar} \\
v_1 &= 30 \text{ m}^3 & v_2 &= ? \\
T_1 &= 273^\circ + 25^\circ & T_2 &= 273^\circ + 0 \\
&= 298 \text{ K} & &= 273 \text{ K}
\end{aligned}
$$

$$\frac{p_1 v_1}{T_1} = \frac{p_2 v_2}{T_2}$$

$$\frac{7 \times 30}{298} = \frac{1{\cdot}01\, v_2}{273}$$

$$v_2 = \frac{7 \times 30 \times 273}{298 \times 1{\cdot}01} = 190 \text{ m}^3.$$

Cylinder A contains 190 m³ at s.t.p.

Cylinder B

$$
\begin{aligned}
p_1 &= 5 \text{ bar} & p_2 &= 1{\cdot}01 \text{ bar} \\
v_1 &= 35 \text{ m}^3 & v_2 &= ? \\
T_1 &= 273^\circ + 20^\circ & T_2 &= 273^\circ + 0^\circ \\
&= 293 \text{ K} & &= 273 \text{ K}
\end{aligned}
$$

$$\frac{p_1 v_1}{T_1} = \frac{p_2 v_2}{T_2}$$

$$\frac{5 \times 35}{293} = \frac{1{\cdot}01\, v_2}{273}$$

$$v_2 = \frac{5 \times 35 \times 273}{293 \times 1{\cdot}01} = 162 \text{ m}^3.$$

Cylinder B contains 162 m³ at s.t.p.

Cylinder A, although smaller than cylinder B, contains more gas than cylinder B.

Example 18.3

One kg of gas at 0°C and 1·01 bar pressure has a volume of 0·5 m³. What will be the mass of the gas which fills a container 3 m diameter and 10 m long at a pressure of 10·1 bar and a temperature of 20°C?

$$\text{Volume of container} = \frac{\pi}{4} \times 3^2 \times 10$$
$$= 70{\cdot}8 \text{ m}^3.$$

We have 70·8 m³ of gas at 10·1 bar pressure and 20°C temperature. We require to know the volume which this gas will occupy at 0°C and 1·01 bar pressure.

$$p_1 = 10{\cdot}1 \text{ bar} \qquad p_2 = 1{\cdot}01 \text{ bar}$$
$$v_1 = 70{\cdot}8 \text{ m}^3 \qquad v_2 = v_2 \text{ m}^3$$
$$T_1 = 20^\circ + 273^\circ = 293 \text{ K} \qquad T_2 = 0^\circ + 273^\circ = 273 \text{ K}$$

$$\frac{p_1 v_1}{T_1} = \frac{p_2 v_2}{T_2}$$
$$\frac{10{\cdot}1 \times 70{\cdot}8}{293} = \frac{1{\cdot}01\, v_2}{273}$$
$$v_2 = \frac{10{\cdot}1 \times 70{\cdot}8 \times 273}{293 \times 1{\cdot}01}$$
$$v_2 = 660 \text{ m}^3 \text{ at } 0^\circ\text{C and } 1{\cdot}01 \text{ bar.}$$

Under these conditions, 0·5 m³ of the gas has a mass of 1 kg. Therefore, 660 m³ have a mass of 660/0·5 = 1320 kg.

Mass of gas in container, 1320 kg.

Example 18.4

A quantity of gas occupies 2 m³ at a pressure of 1 bar absolute and a temperature of 65°C. If the value of R for the gas is 500 J/kg K, determine the mass of gas present.

$$p = 1 \text{ bar} = 100 \text{ kN/m}^2 = 100 \times 1000 \text{ N/m}^2 \qquad v = 2 \text{ m}^2$$
$$R = 500 \text{ J/kg K} = 500 \text{ Nm/kg K} \qquad T = 65^\circ\text{C} = 65 + 273 = 338 \text{ K}$$

$$pv = mRT$$
$$m = \frac{100 \times 1000 \times 2}{500 \times 338}$$
$$= 1{\cdot}18 \text{ kg.}$$

The mass of the gas present is 1·18 kg.

Example 18.5

Two vessels, A, volume 0·8 m³, and B, volume 0·6 m³, are connected by a valve. After the valve is closed, the vessels are charged with a gas at pressures of 5 bar in A and 3 bar in B. The valve is opened and the temperature remains constant while the gas is allowed to mix. What will be the final pressure in the vessels?

If the gas in A were allowed to occupy both A and B (the gas in B considered to have been removed) then let its pressure be p_A.

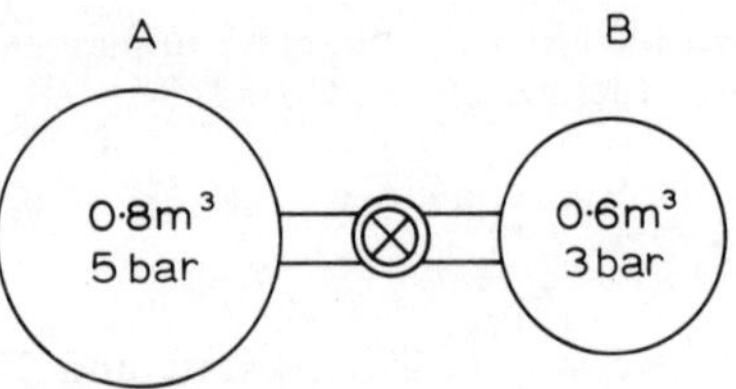

Figure 18.5

Applying Boyle's Law, and noting that the combined volume of A and B is 1·4 m³, we have

$$p_A \times 1{\cdot}4 = 5 \times 0{\cdot}8$$
$$p_A = 2{\cdot}86 \text{ bar.}$$

This is the partial pressure of gas in A.

Similarly, if the gas in B were allowed to occupy both A and B (the gas in A considered to have been removed) let its pressure be p_B, then

$$p_B \times 1{\cdot}4 = 3 \times 0{\cdot}6$$
$$p_B = 1{\cdot}29 \text{ bar.}$$

This is the partial pressure of gas in B.

The final pressure p in the vessels under the conditions stated in the question will be equal to the sum of the partial pressures p_A and p_B

$$\therefore \quad p = 2{\cdot}86 + 1{\cdot}29 = 4{\cdot}15 \text{ bar.}$$

Final pressure in vessels is 4·15 bar (415 kN/m²).

Example 18.6

How much heat would be required to raise the temperature of 50 g of hydrogen by 40°C (*a*) at constant pressure, and (*b*) at constant volume? How much work is done in the first case?

Specific heat capacity at constant pressure = 14·22 kJ/kg K

$\therefore$ Heat required = mass $\times c_p \times$ temperature rise at constant pressure

$$= \frac{50}{1000} \times 14{\cdot}22 \times 40$$
$$= 28{\cdot}44 \text{ kJ.}$$

Specific heat capacity at constant volume = 10·06 kJ/kg K

Heat required at constant volume = mass $\times c_v \times$ temperature rise

$$= \frac{50}{1000} \times 10{\cdot}06 \times 40$$
$$= 20{\cdot}12 \text{ kJ.}$$

Work done when raising temperature at constant pressure

$$= 28{\cdot}44 - 20{\cdot}12$$
$$= 8{\cdot}32 \text{ kJ.}$$

Heat required at constant pressure is 28·44 kJ, at constant volume is 20·12 kJ. Work done in first case is 8·32 kJ.

Example 18.7

How much heat is required to raise the temperature of 1 kmol of hydrogen 1°C (*a*) at constant pressure, and (*b*) at constant volume? How much work is done in (*a*)?

For hydrogen, $c_p = 14{\cdot}22, \quad c_v = 10{\cdot}06.$
The relative molecular mass of hydrogen is 2,
∴ 1 kmol of hydrogen contains a mass of 2 kg.

(*a*) at constant pressure

$$\begin{aligned} \text{Heat required} &= \text{mass} \times c_p \times \text{temperature rise} \\ &= 2 \times 14{\cdot}22 \times 1 \\ &= 28{\cdot}44 \text{ kJ.} \end{aligned}$$

(*b*) at constant volume

$$\begin{aligned} \text{Heat required} &= \text{mass} \times c_v \times \text{temperature rise} \\ &= 2 \times 10{\cdot}06 \times 1 \\ &= 20{\cdot}12 \text{ kJ.} \end{aligned}$$

$$\begin{aligned} \text{Work done at constant pressure} &= 28{\cdot}44 - 20{\cdot}12 \\ &= 8{\cdot}32 \text{ kJ.} \end{aligned}$$

Heat required at constant pressure is 28·44 kJ, at constant volume is 20·12 kJ. Work done in first case is 8·32 kJ.

Example 18.8

Determine the work done when 1 kmol of nitrogen has its temperature raised by 1°C at constant pressure.

For nitrogen, $c_p = 1{\cdot}047, \quad c_v = 0{\cdot}75$
The relative molecular mass of nitrogen is 28,
∴ 1 kmol of nitrogen contains a mass of 28 kilograms.

$$\begin{aligned} \text{Heat required at constant pressure} &= \text{mass} \times c_p \times \text{temperature rise} \\ &= 28 \times 1{\cdot}047 \times 1 \\ &= 29{\cdot}32 \text{ kJ.} \end{aligned}$$

$$\begin{aligned} \text{Heat required at constant volume} &= \text{mass} \times c_v \times \text{temperature rise} \\ &= 28 \times 0{\cdot}75 \times 1 \\ &= 21 \text{ kJ.} \end{aligned}$$

$$\begin{aligned} \text{Work done} &= \text{heat required at constant pressure} - \text{heat required at constant volume} \\ &= 29{\cdot}32 - 21 \\ &= 8{\cdot}32 \text{ kJ.} \end{aligned}$$

Work done at constant pressure is 8·32 kJ.

Note that this is the same quantity as in example 18.7. Note also that this figure of 8·32 kJ/kmol of gas for a temperature rise of 1 K is equal to the universal gas constant, R_0.

In fact, the amount of work done when the temperature of 1 kmol of any gas is raised by 1 K at constant pressure is always equal to the universal gas constant.

$$m(c_p - c_v) = 8{\cdot}32 \text{ kJ}$$

where the mass is 1 kmol of the gas.

Example 18.9

Given that the specific heat capacities of carbon monoxide are

$$c_p = 1{\cdot}04 \quad \text{and} \quad c_v = 0{\cdot}743,$$

calculate the relative molecular mass of the gas.

$$m(c_p - c_v) = R_0$$

where m is the mass of gas expressed in kmol, and R_0 is the universal gas constant 8·32 kJ/kg K.

$$m(1{\cdot}04 - 0{\cdot}743) = 8{\cdot}32$$
$$m = 28$$

i.e. 1 kmol of carbon monoxide has a mass of 28 kg.

By definition, the kmol has a mass in kg equal to the relative molecular mass of the gas.

∴ **Relative molecular mass of carbon monoxide is 28.**

EXERCISE 18

1 A cylinder contains 10 m³ of gas at atmospheric pressure and 117°C. Find its volume at 27°C when the pressure is 2·75 bar absolute.

2 The volume of a gas at s.t.p. is 80 m³. What will its volume be at 4 bar absolute and 30°C?

3 One gram of nitrogen occupies 860 cm³ at 20°C and a pressure of 76 cm of mercury. To what temperature must it be raised to occupy 3020 cm³ at a pressure of 57 cm of mercury?

4 An air compressor is to compress 200 m³ of air at normal atmospheric pressure of 1·01 bar into a receiver having a capacity of 15 m³. Determine the resulting pressure in the receiver after the temperature has cooled down to that of the atmosphere.

5 If the air in question 4 is further compressed until the pressure is 14 bar, determine the new volume of the air if no change in temperature takes place.

6 One m³ of air at 0°C and a pressure of 1 bar has a mass of 1·29 kg. Calculate the mass of air in a 50 m³ container in which the pressure is 7 bar absolute and the temperature 40°C.

7 Four m³ of air at 35 bar absolute and 30°C expands to 8 m³ when the pressure is 7 bar absolute. Calculate the change in temperature. [N.C.T.E.C.]

8 Compressed air at 9 bar absolute and 30°C passes through a heater which raises its temperature to 90°C. If the pressure falls to 8 bar absolute, find the percentage change in the volume of the air. [N.C.T.E.C.]

9 A certain gas occupies a volume of 31 m³ at a temperature of 150°C, and when heated at constant pressure occupies 49 m³ at 400°C. Assuming that the gas is heated and cooled in accordance with the law of Charles, draw a graph of temperature-volume, and estimate (*a*) the volume at 0°C, (*b*) the temperature at which the volume would be zero (assuming the gas remains a gas at this temperature) and (*c*) the change in volume per degree change in temperature expressed as a ratio of the volume at 0°C.

(All temperatures are measured on the ordinary Celsius scale.) [N.C.T.E.C.]

10 What do you understand by the terms 'absolute pressure' and 'absolute temperature'?

State Boyle's and Charles's laws, and show that they may be expressed in the form of a single equation.

Find the volume of air at 1 bar absolute and 50°C which is at 5 bar absolute and 64°C when compressed into a reservoir of 750 m³ capacity. [N.C.T.E.C.]

11 The capacity of an air receiver is 150 m³, and it contains air at a pressure of 8 bar absolute and a temperature of 25°C. If 1 m³ of air at 0°C and a pressure of 1 bar has a mass of 1·29 kg, calculate the mass of air in the receiver. [U.E.I.]

12 In the compression stroke of a diesel engine, 0·1 m³ of air at a pressure of 0·9 bar absolute and a temperature of 45°C is compressed to a pressure of 30 bar absolute and a temperature of 690°C. The compression follows the combined laws of Boyle and Charles. Calculate the volume of the air at the end of the compression. [U.E.I.]

13 At what temperature will 20 kg of air, at a pressure of 1·7 bar absolute, fill a volume of 100 m³, given that 1 kg of air at 1 bar absolute and 0°C has a volume of 0·776 m³? [U.E.I.]

14 One kg of air at 0°C and atmospheric pressure (1·01 bar) has a volume of 0·776 m³. What mass of air will fill a receiver 50 cm in diameter, 2 m long, at a pressure of 14 bar absolute and a temperature of 15°C? [U.E.I.]

15 A room having a volume of 10 000 m³ contains air at 15°C and 1 bar absolute. The temperature is raised to 21°C, the pressure remaining constant. What volume of air measured at 1 bar absolute and 15°C must leave the room?

If the air pressure was 1·1 bar absolute, and temperature remained at 21°C, how much air, measured at 21°C and 1 bar absolute, must you add? (Neglect changes in volume of the room.) [U.L.C.I.]

16 The air in a cylinder has a volume of 6 m³, a pressure of 1·4 bar absolute and temperature of 12°C. The air is compressed at constant temperature until its pressure is 5·5 bar absolute; what is its volume? If the air at 5·5 bar and 12°C is heated at constant volume until its pressure is 6·2 bar absolute, what is its temperature in °C? Neglect changes in volume of the cylinder. [U.L.C.I.]

17 Two cylinders A and B contain compressed air. Cylinder A has a volume of 3 m³, a pressure of 10 bar absolute and a temperature of 15°C; cylinder B has a capacity of 4 m³, a pressure of 14 bar absolute and a temperature of 15°C. If the two cylinders are connected together and the valves opened, what will be their pressure? If the cylinders now have their temperature reduced to 4°C, what would be the common pressure? Neglect change in volume of the cylinders. [U.L.C.I.]

18 Explain what is meant by: (*a*) absolute temperature, and (*b*) absolute pressure.

A quantity of gas occupies a volume of 2 m³ at an absolute pressure of 14 bar and a temperature of 25°C. It is allowed to expand until the pressure has fallen to 8 bar absolute, the temperature also falling to 20°C. What will be the new volume?

If the temperature then increases to 40° C without change in volume, what will be the new pressure? [N.C.T.E.C.]

19 (*a*) A compression chamber has a volume of 60 m³.

When atmospheric pressure is 1 bar with a temperature of 15°C, determine the volume of air at the same conditions to be pumped into the chamber to raise the pressure inside to 7 bar absolute without change in temperature.

(*b*) If the operation is carried out hurriedly so that a temperature of 65°C is reached during the process, find the final pressure when the temperature again falls to 15°C. [N.C.T.E.C.]

20 A quantity of air occupies 0·5 m³ at a pressure of 0·9 bar absolute and temperature 90°C. It is compressed to a volume of 0·03 m³ when the pressure is 27·5 bar absolute. Determine (*a*) the mass of air present and (*b*) the temperature at the end of compression. State clearly the meaning of each symbol used in your solution.

(R for air = 287 J/kg K)

21 How much heat is required to raise the temperature of 1 kmol of oxygen by 40 K (*a*) at constant pressure, and (*b*) at constant volume? How much work is done in (*a*)?

22 Three cubic metres of gas, relative molecular mass 71, are at a pressure of 0·2 bar and a temperature of 20°C. Calculate the mass of the gas.

23 Ten kilograms of helium are contained in a vessel at a pressure of 30 bar and a temperature of 16°C. The gas is heated at constant pressure from a volume of 2 m³ to one of 2·2 m³. Determine the final temperature and the relative molecular mass of the gas.

24 Two containers A and B whose volumes are 2 m³ and 3 m³ are connected together by a control valve and pipe. A contains a gas at a pressure of 8 bar and temperature 20°C while the gas in B has a pressure of 16 bar and a temperature 20°C. The valve is opened and the gases are allowed to mix. What will be the final pressure if the temperature remains at 20°C?

Chapter Nineteen
Combustion

Chemical symbols

Letters are used to represent chemical elements, and numbers are used as suffixes to indicate the number of atoms of the element in the molecule.

For example, O is the symbol for oxygen but, since a molecule of oxygen contains two atoms, oxygen is represented by O_2.

$4O_2$ represents four molecules of oxygen.

Since a molecule of carbon (symbol C) contains only one atom, the molecule of carbon is represented by C.

4C represents four molecules of carbon.

In the same way, S represents a molecule of sulphur and H_2 a molecule of hydrogen.

Combustion

A considerable amount of energy is released when a fuel burns or when combustion takes place. The combustion might be the simple burning of a piece of paper in air, or the more involved combustion of liquid hydrogen with liquid oxygen.

Two essentials are required for combustion:

(*a*) the substance must be capable of burning, and
(*b*) there must be a supply of gas which will support the combustion.

Substances which burn readily are generally known as fuels, and include wood, coal, petrol and coal gas amongst the more common varieties. Oxygen is the most common support for combustion and can be used in gaseous form, as part of the atmosphere, or as a liquid. Oxygen is also present in some fuels, although oxygen from an external supply is always needed for complete combustion.

Once a fuel has had its temperature raised to ignition temperature, it will continue to burn provided a supply of oxygen is available. An adequate supply of oxygen is essential if the maximum amount of heat is to be obtained from the fuel.

Solid fuels are usually burnt in a boiler supplied with a quantity of air, often under pressure (forced draught) and the heat released is transferred to steam which can be used to drive an engine or operate a process.

Liquid and gaseous fuels can be inserted into a cylinder with a quantity of air and the mixture exploded by an electric spark. Combustion takes place internally in the cylinder as distinct from external combustion as with the boiler. The internal combustion engine operates on this principle with fuel and air being inserted in a vaporised state into the cylinders.

Combustion is a chemical process which can be expressed by means of chemical equations, whereby we can write down in terms of letters and numbers the reaction which takes place and the products which are formed when molecules of different substances combine. The most common substance in fuels is carbon.

Substance	Relative atomic mass	Relative molecular mass
hydrogen	1	2
carbon	12	12
nitrogen	14	28
oxygen	16	32
sulphur	32	32

Now carbon combines with oxygen to produce either carbon monoxide or carbon dioxide, depending on how much oxygen is present. If insufficient oxygen is supplied, carbon monoxide is formed; otherwise the product is carbon dioxide.

The equation can be written:

$$\text{Carbon} + \text{oxygen} = \text{carbon dioxide}$$

Using the symbol C for carbon and O_2 for oxygen, we can write:

$$C + O_2 = CO_2$$

where CO_2 is the symbol for carbon dioxide.

One molecule of carbon combines with one molecule of oxygen to form one molecule of carbon dioxide.

If insufficient oxygen is present then we write:

$$\text{Carbon} + \text{oxygen} = \text{carbon monoxide}$$
$$2C + O_2 = 2CO$$

where CO is the symbol for carbon monoxide.

Two molecules of carbon combine with one molecule of oxygen to form two molecules of carbon monoxide.

Note the oxygen symbol is written as O_2.

Look again at the equation for carbon dioxide:

$$C + O_2 = CO_2.$$

One molecule of carbon requires one molecule of oxygen to enable it to burn completely, or 1 kmol of carbon requires 1 kmol of oxygen to enable it to burn completely.

1 kmol of carbon has a mass of 12 kg
1 kmol of oxygen has a mass of 32 kg.

Hence, 12 kg of carbon combines with 32 kg of oxygen to form 44 kg of carbon dioxide,

or 1 kg of carbon combines with $2\frac{2}{3}$ kg of oxygen to form $3\frac{2}{3}$ kg carbon dioxide.

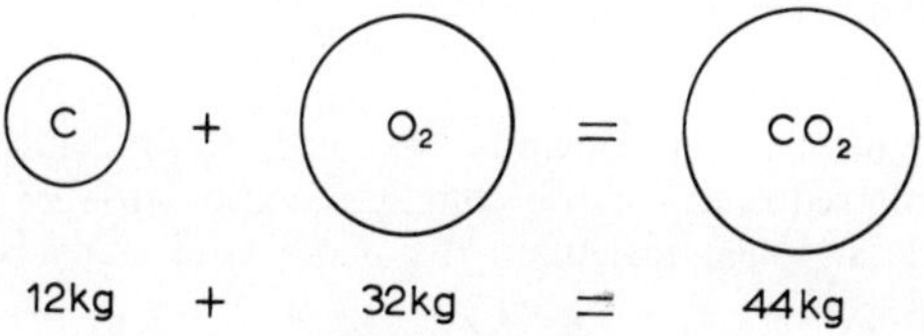

$C + O_2 = CO_2$

12kg + 32kg = 44kg

Figure 19.1

Now air is made up of oxygen and nitrogen, in the following proportions by mass or volume. By mass, air contains 23 per cent. oxygen and 77 per cent. nitrogen. By volume, air contains 21 per cent. oxygen and 79 per cent. nitrogen.

The difference between the proportions is due to the fact that the density of oxygen is different from that of nitrogen.

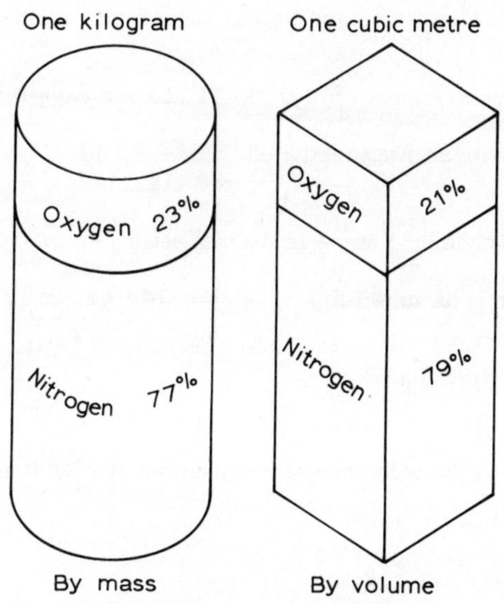

Figure 19.2 *Composition of Air*

Normally we use the mass proportion when dealing with solid fuels and the volume proportion when dealing with gaseous fuels.

So $2\frac{2}{3}$ kg of oxygen are contained in $2\frac{2}{3} \times \frac{100}{23}$ kg of air, i.e. 11·6 kg of air.

Therefore 1 kg of carbon requires 11·6 kg of air for complete combustion.

It is normal to arrange for an air supply in excess of this to ensure that the combustion is efficiently carried out. Up to 5 per cent excess air is often used, dependent upon the fuel and the design of the combustion chamber.

Example 19.1

Petrol contains 86 per cent. carbon and 14 per cent. hydrogen by mass. Calculate the quantity of air required for the complete combustion of 1 kg of the fuel. If 20 kg of air is supplied, calculate the percentage analysis by mass of the products of combustion.

Equation for carbon:
$$C + O_2 = CO_2$$
$$12 + 32 = 44$$

1 kg of carbon requires $\frac{32}{12}$ kg of oxygen for complete combustion,

0·86 kg of carbon requires $\frac{32}{12} \times 0{\cdot}86 = 2{\cdot}29$ kg of oxygen.

Equation for hydrogen:
$$2H_2 + O_2 = 2H_2O$$
$$4 + 32 = 36.$$

1 kg of hydrogen requires $\frac{32}{4}$ kg of oxygen for complete combustion,

0·14 kg of hydrogen requires $\frac{32}{4} \times 0{\cdot}14 = 1{\cdot}12$ kg of oxygen.

$$\text{Quantity of oxygen required} = 2{\cdot}29 + 1{\cdot}12 = 3{\cdot}41 \text{ kg.}$$

1 kg of oxygen is contained in $\frac{100}{23}$ or 4·35 kg of air

3·41 kg of oxygen is contained in $4{\cdot}35 \times 3{\cdot}41 = 14{\cdot}8$ kg of air.

1 kg of air occupies 0·82 m^3 at 15°C and a pressure of 1 bar.
Hence volume of air required is

$$14{\cdot}8 \times 0{\cdot}82 = 12{\cdot}15 \text{ m}^3.$$

Quantity of air required for complete combustion of 1 kg of petrol is 12·15 m^3.

Products of combustion

1 kg of carbon produces $\frac{44}{12}$ kg of CO_2

0·86 kg of carbon produces $\frac{44}{12} \times 0{\cdot}86 = 3{\cdot}15$ kg CO_2.

1 kg of hydrogen produces $\frac{36}{4} = 9$ kg of H_2O

0·14 kg of hydrogen produces $9 \times 0{\cdot}14 = 1{\cdot}26$ kg H_2O.

20 kg of air contains $20 \times 0{\cdot}77$ kg $= 15{\cdot}4$ kg nitrogen.

Excess air is $20 - 14{\cdot}8 = 5{\cdot}2$ kg.

This contains $5{\cdot}2 \times 0{\cdot}23$ kg $= 1{\cdot}19$ kg oxygen.

The products of combustion are:

	kg	%
carbon dioxide CO_2	3·15	15·0
water H_2O	1·26	6·0
nitrogen N	15·40	73·5
oxygen O_2	1·19	6·5
	21·00 kg	100·00

Note that this is equal to 20 kg of air plus 1 kg of fuel.

$$\% \text{ carbon dioxide} = \frac{3{\cdot}15}{21} \times 100$$

$$= 15 \text{ per cent.}$$

and similarly for the other products.

Calorific value of fuel

The purpose of burning fuel is to release heat energy from it. The questions which arise are: how much heat energy do we get? how does one fuel compare with another?

The amount of heat energy released when a unit quantity of fuel is burnt completely is called the *calorific value* of the fuel. It is measured in kilojoules per cubic metre in the case of gaseous fuels or kilojoules per kilogram in the case of solid or liquid fuels. The question of the distinction between the higher and lower calorific values of a fuel will be under consideration when you carry out further work in the subject of applied heat.

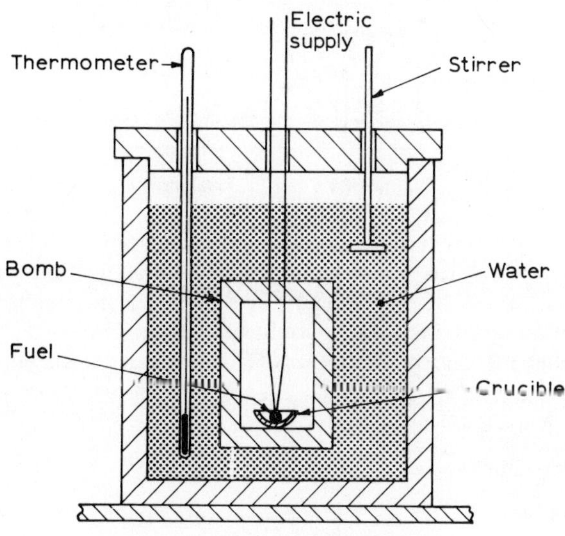

Figure 19.3 *Calorific Value of Solid*

The experimental determination of calorific values is carried out by means of suitable calorimeters. A known quantity of the fuel is burnt in a container surrounded by water, and the temperature rise of the water is observed.

Heat given out by fuel = heat received by water + heat received by calorimeter

Since the heat received by the water is equal to the mass of water × specific heat capacity of water × temperature rise, and that received by the metal calorimeters is equal to mass of calorimeter × specific heat capacity of the metal × temperature rise, it is possible to calculate the amount of heat given out by the fuel.

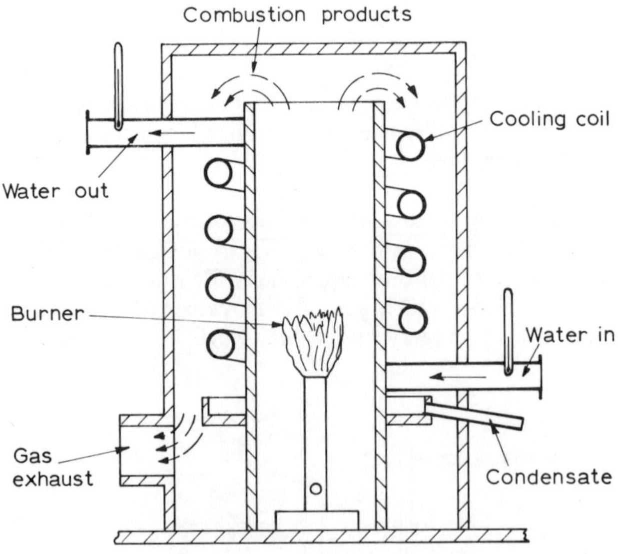

Figure 19.4 *Calorific Value of Gas*

In order to avoid a repetition of the calculation involving the heat received by the calorimeter, it is usual to use an expression known as the *water equivalent* of the calorimeter. This indicates the quantity of water which would require the same amount of heat to raise its temperature by one degree as the calorimeter requires for the same temperature rise.

Water equivalent × specific heat capacity of water = mass of calorimeter × specific heat capacity of calorimeter.

Water equivalent (kg) =

$$\frac{\text{mass of calorimeter (kg)} \times \text{specific heat capacity of calorimeter}}{\text{specific heat capacity of water}}$$

In the case of a copper calorimeter, the ratio of the specific heat capacities is 0·1, which means that the water equivalent of the calorimeter is a mass of water equal to 0·1 of the mass of the copper calorimeter.

The water equivalent is then added to the mass of water in the calorimeter and the calculation carried out as though this new quantity of water was the only substance to receive heat in the experiment.

Incidentally, a reminder might be given here that the specific heat capacity of water (at 15°C) is 4·19 kJ/kg K.

Example 19.2

One gram of coal was burned in a calorimeter containing 2 kg of water and the temperature rise was observed to be 3·5 K. The water equivalent of the calorimeter was 0·3 kg. Calculate the calorific value of the fuel, assuming that there were no heat losses.

$$\text{Equivalent mass of water} = 2 + 0{\cdot}3 = 2{\cdot}3 \text{ kg.}$$

Heat required to raise temperature of this mass of water by 3·5 K is given by

$$\text{mass} \times \text{specific heat capacity} \times \text{temperature rise}$$

$$= 2{\cdot}3 \times 4{\cdot}19 \times 3{\cdot}5 = 33{\cdot}7 \text{ kJ.}$$

This heat is released when 1 gram of the fuel is burned. When 1 kg of fuel is burned, heat released will be 33 700 kJ or 33·7 MJ.

Hence, calorific value of coal is 33·7 MJ per kg.

Example 19.3

The following results were recorded during an experiment to determine the calorific value of North Sea gas.

quantity of water collected	1 kg
inlet temperature	15°C
outlet temperature	33·9°C
gas consumed	2000 cm^3
gas pressure	9 cm water above that of atmosphere
gas temperature	17°C
atmospheric pressure	1·05 bar

Calculate the calorific value of the gas in megajoules per metre^3 at 15°C and 1 bar, i.e. in MJ per standard metre^3.

$$\text{Heat released by gas} = \text{heat received by water}$$

$$\text{Heat received by water} = \text{mass} \times \text{temperature rise} \times \text{specific heat capacity}$$
$$= 1 \times (33{\cdot}9 - 15) \times 4{\cdot}19$$
$$= 79{\cdot}2 \text{ kJ.}$$

$$\text{Gas pressure} = 9 \text{ cm } H_2O \text{ gauge}$$
$$= 0{\cdot}09 \times 9{\cdot}81 \times 1000 \text{ N/m}^2$$
$$= 0{\cdot}88 \text{ kN/m}^2$$
$$= 0{\cdot}0088 \text{ bar gauge.}$$

$$\text{Absolute gas pressure} = 0{\cdot}0088 + 1{\cdot}05$$
$$= 1{\cdot}059 \text{ bar.}$$

Volume of gas at 15°C and pressure of 1 bar is given by

$$\frac{pv}{T} = \frac{p_g v_g}{T_g}$$

$$v = \frac{p_g v_g}{T_g} \times \frac{T}{p}$$

$$= \frac{1{\cdot}059 \times 2000 \times (273 + 15)}{(273 + 17) \times 1}$$

$$= 2103{\cdot}3 \text{ cm}^3$$

$$\therefore \quad \text{Calorific value of gas} = \frac{\text{heat released by gas}}{\text{volume}}$$

$$= \frac{79{\cdot}2 \times 100 \times 100 \times 100}{2103{\cdot}2}$$

$$= 37\,650 \text{ kJ/m}^3$$
$$= 37{\cdot}65 \text{ MJ/m}^3.$$

Calorific value of North Sea gas is 37·65 MJ per standard metre^3 (15° C and 1 bar pressure).

EXERCISE 19

1 The analysis of fuel supplied to a boiler is 83 per cent. carbon, 8 per cent. hydrogen, 2 per cent. oxygen and the remainder is incombustible matter. Calculate the theoretical mass of air required for combustion of 1 kg of the fuel. (Note that some oxygen is already in the fuel.)

2 A fuel has the following composition by mass: carbon 81 per cent., hydrogen 8 per cent., remainder ash. Calculate the theoretical mass of air necessary for complete combustion of 1 kg of the fuel.

Calculate also the mass analysis of the post-combustion mixture. (Relative atomic masses C: 12, O: 16, H: 1. Assume air contains 23·3 per cent. oxygen by mass.) [U.L.C.I.]

3 2000 g of water is in a copper calorimeter and has its temperature raised by 5 K when 1 g of fuel oil is burned. The water equivalent of the calorimeter is 400 g. What is the calorific value of the oil?

4 Give a simple sketch showing the construction of a gas calorimeter with a line diagram of the water and gas circuits.

The following results were recorded during an experiment using a gas calorimeter to determine the energy of combustion (calorific value) of coal gas:

mass of water collected	800 g
water inlet temperature	14°C
water outlet temperature	28°C
gas consumed	3000 cm^3
gas pressure	9 cm of water above atmospheric

gas temperature	18°C
barometric pressure	780 mm Hg

Determine the energy of combustion of the gas in MJ per standard cubic metre.

(A standard cubic metre is measured at 15°C and 760 mm Hg.)

[U.L.C.I.]

Appendix 1

Methods of determining area under curve

(1) *Counting squares.* Count the number of whole squares. If the curve passes through a square, count the square in the total, provided that more than half the area of the square is under the curve. If less than half the area is under the curve, then neglect the square.

(2) *Mid-ordinate method.* Divide the total base into a suitable number of equal parts and erect ordinates at the mid-point of the base of each strip. Each strip is then considered to have an area equal to the base × mid-ordinate length. Since each strip has the same width of base, the total area is equal to the sum of the mid-ordinates multiplied by the width of one strip.

(3) *Simpson's rule.* The most accurate method of the three. Divide the base into an even number of equal parts, and erect ordinates at each point of division. Number the first ordinate 1, the next 2, and so on, finishing with an odd number. The area of the figure is given by

$$\text{Area} = \frac{\text{width of one strip}}{3} \times \text{sum of} \begin{bmatrix} \text{first} + \text{last ordinate} \\ 2 \times \text{sum of odd ordinates} \\ 4 \times \text{sum of even ordinates} \end{bmatrix}$$

The first and the last ordinates must not be included in the sum of the odd ordinates.

Appendix 2
Angles greater than 90°

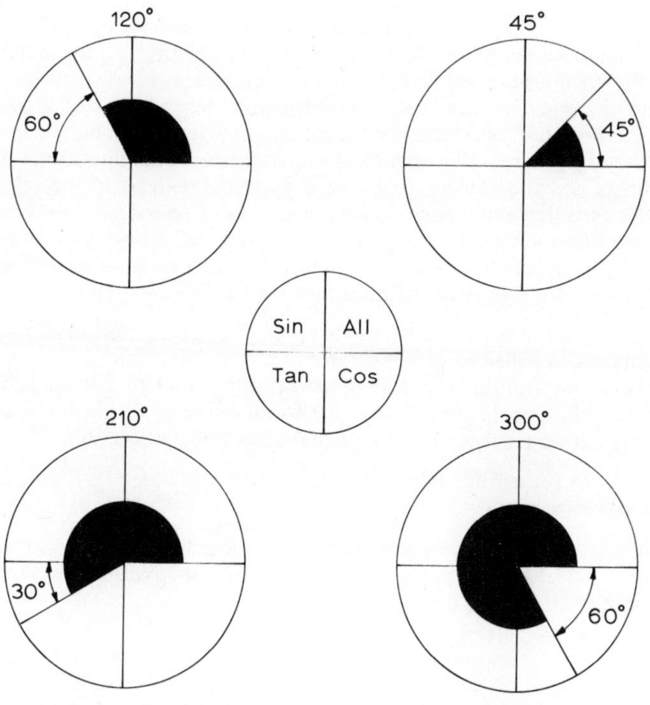

Figure A2.1

Fig. A2.1 is included to help you remember some facts which you will have received in the maths class. We have had to use angles greater than 90° in some of the examples in this book. Remember to convert angles greater than 90° into angles less than 90° by the method illustrated.

Remember, also, that the sin, cos and tan of all angles between 0° and 90° are all positive.
The sin of angles between 90° and 180° is positive.
The tan of angles between 180° and 270° is positive.
The cos of angles between 270° and 360° is positive.
Values for other angles are all negative.

Appendix 3
Sir Isaac Newton and the Laws of Motion

Sir Isaac Newton, who was born in Grantham, Lincolnshire, in 1642 and died in 1727, made extremely far-reaching discoveries relating to the differential calculus, the composition of light, and gravitational attraction while only a young man of twenty-four. In 1685 he wrote his great work *Philosophiae Naturalis Principia Mathematica*, once described as the supreme feat of the human intellect, in which amongst many other remarkable investigations relating to the motion of the planets and the comets, to physical mathematics and to the science of fluid movements he expounded what are generally referred to as Newton's Laws of Motion. Although the principles behind these laws had been used previously by scientists like Galileo, Newton was the first man to record them in formal shape. He expressed them thus.

(1) Every body continues in its state of rest, or of uniform motion in a straight line, unless compelled by external force to change that state.
(2) Change of momentum per unit time is proportional to the applied force and takes place in the direction of the straight line in which the force acts.
(3) To every action there is always an equal and opposite reaction.

In addition to these three laws there is a fourth known as Newton's Law of Gravitation, which states:

Every particle of matter attracts every other particle of matter with a force which varies directly as the product of the masses of the particles and inversely as the square of the distance between them.

Answers

EXERCISE 2

1 (*a*) 68·95, 53° 37′ S of E (*b*) 38·27, 67° 30′ S of E
2 (*a*) 16·15, 278° 16′ (*b*) 27·48, 175° (*c*) 36·05, 303° 41′
3 $103{\cdot}2_{133^\circ\,28'}$
4 (*a*) $30{\cdot}74_{2^\circ\,38'}$ (*b*) $86{\cdot}9_{223^\circ\,18'}$
5 (*a*) 173·2, 100 (*b*) 35·35, 35·35 (*c*) 0, 100 (*d*) −69·28, 40
(*e*) −40, −34·64 (*f*) 50, −86·6
6 83·24 km, 19° 52′ S of E
7 $5{\cdot}44_{180^\circ\,14'}$
8 $5{\cdot}44_{0^\circ\,14'}$
9 216·5 units, 125 units

EXERCISE 3

1 25 m/s
2 10·5 m/s, 190·4 s, 0·047 m/s²
3 53 km/h, 0·33 m/s²
4 12·25 m/s
5 375000 m/s², 0·0008 s
6 213·9 m, 15·4 s
7 52·8 seconds behind time
8 61·9 km/h
9 10·33 s
10 5 m/s
11 56·25 m
12 0·0857 m/s², 11 m/s
13 30 m, 46 s
14 (*a*) 4·44 m/s (*b*) 133·2 m
15 0·038 m/s²
16 (*a*) 0·12 m/s² (*b*) 20·77 km/h, 24·22 km/h
17 1·5 m/s²
18 0·89 m/s², 54·61 km/h, 80·39 km/h, 99·61 km/h
19 (*a*) 0·53 m/s² (*b*) 0·8 m/s² (*c*) 140 m

EXERCISE 4

1 32·4 km/h, N 19° 7′ E
2 3 m/s due E, 1·414 rad/s clockwise
3 3·57 m/s, 52° 29′ anticlockwise from horizontal
4 v_A = 8·66 m/s, 120°, v_C = 4·51 m/s, 196° 6′, 1·25 rad/s clockwise
5 25·05 km/h, S 33° 46′ W
6 17·1 km/h, S 82° 57′ W
7 (*a*) 2·31 m/s (*b*) 2·41 m/s, 28° to line of stroke
(*c*) 7·36 rad/s anticlockwise
8 (*a*) 1·93 m/s (*b*) 2·41 m/s, 36° to line of stroke
(*c*) 24·4 rad/s anticlockwise

9 (*a*) 67·57 rev/min (*b*) (i) 2·12 rad/s clockwise (ii) 7·78 m/s, 316° to line of stroke

EXERCISE 5

1 30·66 m, 24·5 m/s
2 12·74 km, 4·17 s and 97·76 s
3 32·62 km, 163·1 s
4 9·81 m/s, 29·43 m/s; 4·9 m, 44·14 m; 4·52 s, 44·3 m/s
5 (*a*) 2·04 s (*b*) 3·06 s (*c*) 45·87 m
6 5·08 s
7 3028·9 m
8 0·136 m
9 22·37 m/s

EXERCISE 6

1 (*a*) 31·41 rad/s (*b*) 0·526 rad/s (*c*) 477·55 rev/min (*d*) 1719·2 rev/min
2 (*a*) 0·0069 rad/s^2 (*b*) 10·47 rad/s^2
3 27·8 rad/s
4 104·7 m/s
5 10·47 m/s; 27·92 rad/s, 266·7 rev/min
6 448 rev/min; 243 rev/min
7 0·262 rad/s, 750 rev
8 79·6 rev/min, 0·29 m/s^2, 0·58 rad/s^2
9 52·35 rad/s^2, 416·7 rev
10 (*a*) 0·4 m/s^2 (*b*) 177 rev/min, 1·07 rad/s^2
11 (*a*) 1·152 s (*b*) 6·03 m/s^2 (*c*) 12·06 rad/s^2
12 (*a*) 0·026 rad/s^2 (*b*) 480 rev
13 15·7 rad/s, 7·85 m/s
14 (*a*) 229 rev/min (*b*) 2·4 rad/s^2 (*c*) 19·1 rev
15 (*a*) 60 m/s (*b*) 186·6 rev (*c*) 0·75 rad/s^2

EXERCISE 7

1 400 N
2 0·25 m/s^2
3 173·6 s
4 0·033 m/s^2
5 0·28 m/s^2, 4·17 N, 5·83 N
6 2314·6 N
7 (*a*) 5405 N (*b*) 4405 N
8 122·2 N
9 (*a*) 2314·6 N (*b*) 7·2 s
10 8·9 N
11 3·52 N
12 147·8 N
13 19·2 s, 0·72 m/s^2, 1084·8 N
14 (*a*) 1·006 m/s^2 (*b*) 13·81 s (*c*) 95·93 m; 1·99 m/s^2
15 313·3 N, 0·38 m/s^2, 36·15 s, 755 m
16 643 N, 21·6 s
17 0·45 m/s^2, 42·34 s from rest

EXERCISE 8

1 1885·2 N m
2 3820 N m
3 130·9 kW
4 50 kW
5 2500 N m, 21·43 kW
6 552·8 kW

EXERCISE 9

1 112·5 kJ, 109·4 kJ
2 6·25 kN, 15·63 MJ, 15·63 MJ
3 538 J
4 138·1 rev
5 (*a*) 355·17 kJ (*b*) 9·44 kJ
6 (*a*) 1065·5 kJ (*b*) 880·53 kJ
7 408·5 J
8 35·16 kJ, 937·5 N
9 63·76 J, 4·43 m/s, 0·53 m
10 11·43 kJ, 3·81 kN
11 5·39 kJ
12 1·94 MJ, 46·23 kW
13 (*a*) 174·13 kN m (*b*) 56·87 kN m (*c*) 284·4 kW
14 0·14 rad/s, 480 rev, 19·6 kJ, 175·9 m/s
15 120 rev/min, 0·157 rad/s^2, 420 rev, 17·05 N m
16 (*a*) 1·14 kJ (*b*) 1·068 m/s

EXERCISE 10

1 833·3 m/s^2
2 1753·9 m/s^2
3 165·4 rev/min
4 3·7 kN
5 20 m/s or 191 rev/min
6 1·48 kN
7 31·25 kN
8 5·02 kN
9 10·66 kg, 77° to 4·5 kg mass
10 138° 26′ between 10 kg and 7 kg, 71° 24′ between 7 kg and 14 kg, 150° 10′ between 14 kg and 10 kg
11 143° 7′ between casting and 10 kg, 90° between 10 kg and 7 kg, 126° 53′ between 7 kg and casting
12 900 N, 62° 26′ to 10 kg
13 9·9 m/s
14 (*a*) 2302 N (*b*) 52·5 kg
15 1649 N, 10·44 kg, 84° 30′ clockwise from 5 kg
16 841·9 N, 940 N, 743·8 N, 27·3 rev/min

EXERCISE 11

1 (*a*) 240 N m (*b*) 10 m (*c*) 300 N m (*d*) 80% (*e*) 24 N
2 (*a*) 30 N (*b*) 60% (*c*) 40 N
3 (*a*) 12 N (*b*) 75% (*c*) 15 N

4 226·2, 55%
5 3·5, 142·8 N, 178·57 N
6 50%
7 180 N
8 12·5%
9 375 N
10 13·33 rev/min
11 (*a*) 120 (*b*) 166·66 N (*c*) 222·2 N
12 140 rev/min, 80 N
13 (*a*) 512 N (*b*) 16 (*c*) 12·8
14 1·13 m/s
15 (*a*) 480 rev/min (*b*) 257·14 rev/min
16 300 rev/min, 187·6 rev/min, 94·73 rev/min, 59·26 rev/min
17 57·1 N
18 32
19 M.A. 8, V.R. 8; M.A. 5·6, V.R. 8; 70%
20 22·5, 30 N
21 20 rev/min
22 0·51 mm
23 11·4
24 666·7 N
25 6
26 $E = 0{\cdot}125\,W + 0{\cdot}75$
27 50, 48%
28 6, 59%
29 130, 69%
30 7·7 N
31 10 500 N
32 (*a*) 5 (*b*) 125 N (*c*) 4
33 39·5%
34 47·5%
35 10·6 mm, 88·2%
36 703, 23·7 kN
37 960 N
38 419, 48, 11·5%
39 2·22 mm
40 19 N, 29%
41 40, 15·24 N, 12·5 N
42 43·6%
43 26·34 N, 63·8%, 68·5%, 69%
44 1·65 kN
45 178·6 N, 0·224 m/s
46 209·5, 9·5%, 21·2%
47 4·96 N
48 30, 0·78 N, 64·1%
49 (*a*) 0·32 m/s (*b*) 8 kW

EXERCISE 12

1 1·5 cm, 2 cm
2 5·2 cm, 4·6 cm

3 2·66 cm, 5·64 cm
4 2·88 cm, 3·62 cm
5 2·46 cm, 3·5 cm
6 8 cm, 8·25 cm
7 2·39 cm, 2·12 cm
8 2·14 cm, 2·94 cm
9 On centre line, 10 cm from 10 cm edge
10 209·1 mm
11 84·4 mm from short side, 9·4 mm from long side
12 125·6 mm from base, on centre line of solid
13 36·68 cm
14 (*a*) 2 m from one end (*b*) 0·5 m from both legs
(*c*) 0·78 m from short leg, 0·28 m from long leg
15 1·35 m
16 275 cm
17 26° 34′
18 104·6 mm from base
19 (*a*) 6·5 N (*b*) 21 N, 18° to the vertical
20 37·14 mm from 50 mm edge, 12·14 mm from 100 mm edge
21 1·09 m from base, 0·84 m from r.h. edge
22 0·41 m from base, 1·21 m from l.h. edge
23 6·88 cm from circular base
24 6·45 cm dia.
25 1·53 m from base, 0·4 m from back vertical
26 6·36 cm

EXERCISE 13

1 (*a*) 38·5 N, 66° 8′ N of W (*b*) 8·8 N, 54° 37′ S of W
2 1800 N, 1587·4 N inclined at 10° 42′ to the horizontal
3 9414 N, 8077 N at 17° 17′ to the vertical
4 (*a*) 134·6 N (*b*) 175·8 N
5 144·3 N; 520 N at 16° 6′ to the vertical
6 5 N, 8·66 N at 30° to the vertical
7 1777 mm^2
8 (*a*) 800 N, 12° 3′ S of E (*b*) 8 mm^2
9 15·44 N, 36° 34′ to AD approx along diagonal AC
10 10° 46′, 2·36 N in shorter cord, 2·95 N in longer cord
11 4·24 N at 16·5 cm from D acting parallel to CA and outside the square beyond B
12 1·5 kN in the opposite direction to 6 kN,
2·6 kN in the direction perpendicular to 6 kN to the right
13 Wall reaction 0·102 W, ground reaction 1·005 W at 5° 49′ to the vertical
14 12·16 N at 48° 53′ to the vertical, 9·16 N
15 (*a*) 10 N inside square on bisector of AB and DC, 0·7 m from AB
(*b*) 2 N outside square (in direction of 4 N), 0·5 m from AB on bisector of AB and DC
16 0·208, 11° 45′
17 0·192
18 (*a*) 150 N (*b*) 147·6 N
19 0·4

20 100·5 kJ
21 40 N m, 209·5 W
22 575 N

EXERCISE 14

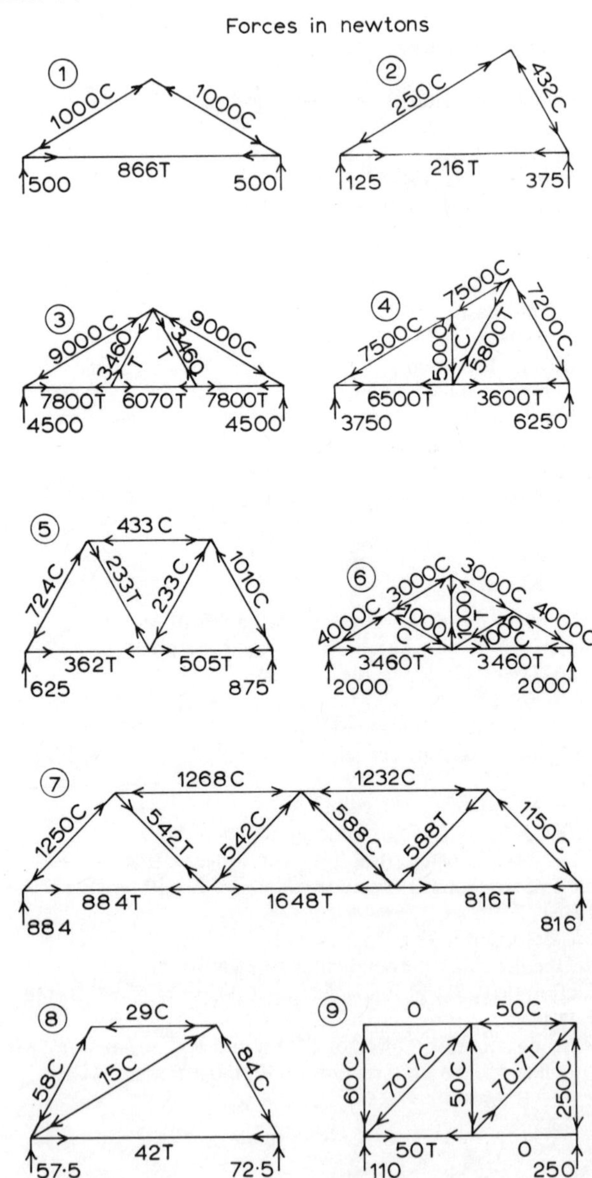

10

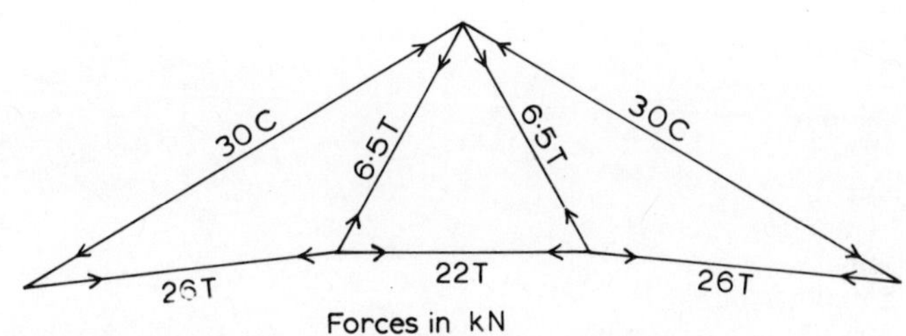

11 (*a*) vertical (*b*) 2·16 kN; 3·3 kN, 49° 4′ to the right of vertical
(*c*) $x=$ 4·33 kN comp, $y=$ 6·25 kN tension

12

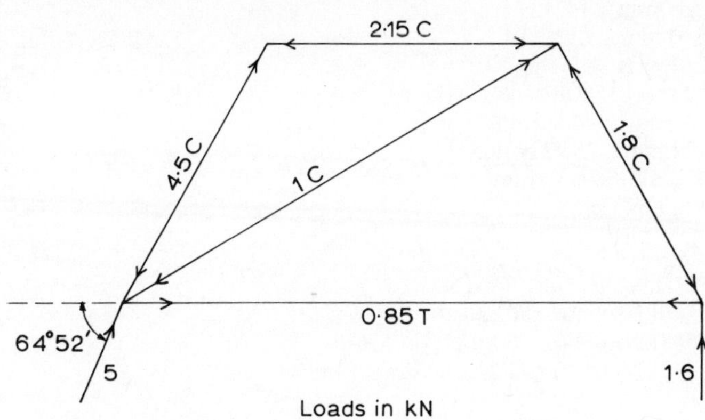

13

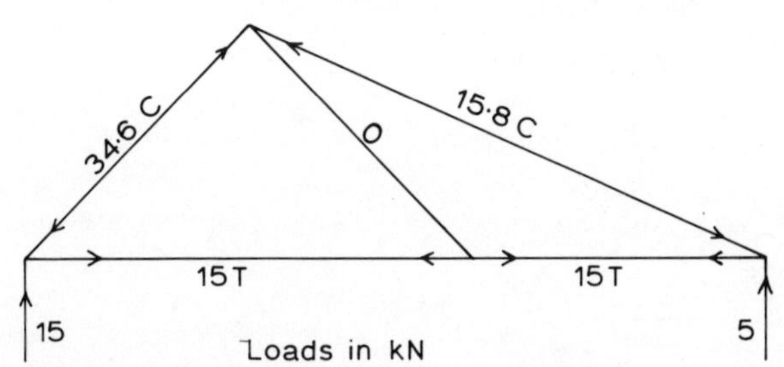

14 PS 14·1 kN comp, SW 8·5 kN comp, RW 8 kN tension
15 AF 9·9 kN comp, BC 8 kN tension, BF 7 kN tension
16 CE 11·3 kN tension, EB 14 kN comp, AB 22 kN comp

EXERCISE 15

1 40 N/mm², 0·1 mm
2 736·5 N, 187·5 kN/mm²
3 35·7 mm, 1·14 mm
4 39 mm, 34 mm

5 0·63 MN, 2000 N/mm²
6 104·2 kN
7 0·41 mm, 0·41 J
8 13·5 J
9 208 kN/mm²
10 314·2 kN
11 132 mm
12 (*a*) 8·1 mm (*b*) 2·76
13 3·98 mm
14 128·2 N/mm², 3·2 mm
15 106 N/mm², 353·7 kN/mm²
16 (*a*) 1·013 J (*b*) 1·8 J (*c*) 36%, 64%
17 (*a*) 37 mm (*b*) 0·0022 mm
18 253·2 N/mm², 7·9 kN/mm², 77·64 kN
19 109·4 N/mm², 147 kN
20 (*b*) 22·1 mm
21 (*b*) 220·9 kN, 1·69 mm
22 (*b*) 32 J (*c*) 15·8 m/s
23 (*a*) 222 mm² (*b*) 0·9 mm
24 213 kN/mm², 285·2 N/mm²
25 (*a*) 67·7 mm (*b*) 3·8 mm
26 198 kN/mm², no

EXERCISE 16

1 (*a*) 78·4 kN/m² (*b*) 7·84 MN
2 (*a*) 2546 kN/m² (*b*) 80 kN
3 (*a*) 5568 kN/m² (*b*) 2800 kN (*c*) 576 m
4 (*b*) 1375 kN/m² (*c*) 42·2 MN
5 39·23 kN/m²
6 12 kW
7 197 kN/m², 26·1 kW
8 19·1 kJ
9 8·11 mm 973·2 kN/m² 3·87 kW
10 1·09 m
11 0·68 bar
12 0·4 bar
13 0·047 bar

EXERCISE 17

1 24 kJ
2 920 J
3 80·5 J
4 32·2 mm
5 1013 W
6 826·1 W
7 1776°C
8 6·32 W

EXERCISE 18

1 2·83 m³
2 22·42 m³

3 498·7°C
4 13·47 bar
5 14·43 m^3
6 393·8 kg
7 Drops to −151·8°C
8 34·7% increase
9 (*a*) 20·2 m^3 (*b*) −280·6°C (*c*) 0·072 per °C
10 3594 m^3
11 1418·1 kg
12 0·0091 m^3
13 2713°C
14 6·65 kg
15 208·3 m^3, 792 m^3
16 1·53 m^3, 48·2°C
17 12·28 bar, 11·81 bar
18 3·44 m^3, 8·54 bar
19 (*a*) 420 m^3 (*b*) 7 bar
20 (*a*) 0·432 kg (*b*) 392·5°C
21 (*a*) 1164·8 kJ (*b*) 832 kJ; 332·8 kJ
22 1·75 kg
23 44·9°C, 4
24 12·8 bar

EXERCISE 19

1 12·3 kg
2 12·02 kg; carbon dioxide 2·97 kg, nitrogen 9·22 kg, water 0·72 kg, ash 0·11 kg
3 50·2 MJ/kg
4 15·1 MJ/m^3

Index